The Automatic Self

The Automatic Self

◆

Transformation & Transcendence Through Brainwave Training

Richard G. Soutar, Ph.D.

iUniverse, Inc.
New York Lincoln Shanghai

The Automatic Self
Transformation & Transcendence Through Brainwave Training

Copyright © 2006 by Richard G. Soutar

All rights reserved. No part of this book may be used or reproduced by any means, graphic, electronic, or mechanical, including photocopying, recording, taping or by any information storage retrieval system without the written permission of the publisher except in the case of brief quotations embodied in critical articles and reviews.

iUniverse books may be ordered through booksellers or by contacting:

iUniverse
2021 Pine Lake Road, Suite 100
Lincoln, NE 68512
www.iuniverse.com
1-800-Authors (1-800-288-4677)

ISBN-13: 978-0-595-38695-6 (pbk)
ISBN-13: 978-0-595-83078-7 (ebk)
ISBN-10: 0-595-38695-4 (pbk)
ISBN-10: 0-595-83078-1 (ebk)

Printed in the United States of America

This book is dedicated to everyone past and present who is devoted to understanding the human condition, increasing human wisdom and reducing human suffering. It is a book I envisioned writing at my present age when I was much younger. Many of the concepts in the book emerged from the intellectual struggles of my youth. At the age of eighteen I stopped reading fiction or anything that was not directly related to the task of understanding the human riddle. As the insights grew, I resolved not to commit them to paper officially until I had experienced a major span of human life. Now, having experienced a wife, children, a professional life and the ups and downs of modern business, I feel more secure in the validity of much of my perspective. It is not a static one. Since it is grounded in science, it has grown with the research. It continues to grow.

As I have grown and developed as an individual, I have gravitated toward the neurosciences and the field of quantitative EEG Topographic Brain Mapping and EEG Biofeedback. I have brought along with me the wisdom I have garnered from years of studying traditional modes of transformation and transcendence both in text and through practice and experience. I have found them progressively less contradictory. It seems evident at this point that those of us who live in modern western culture have difficulty accessing the more ancient methods of transformation and transcendence and that we will need to rely on our technology to do the same job as those older methods once did.

With this in mind, over the past decade and a half I have personally explored and clinically employed the technologies I discuss in this text. The results have astounded my clients as much as myself and other professionals. What still seems like science fiction to me is today a reality. Most of the public is totally unaware of this technology and its potential, but they are slowly becoming aware of it. It has the potential to transform the social order in a manner we have only dreamed of in the past. If we survive as a society long enough to employ it properly, I believe it will transform the future in the way that we have always hoped for as a species. The power of the neuro-technologies and the people who use them is already shaping our current reality and our future.

I would like to extend special thanks to my wife and children who sacrificed their personal time with me so that I would have the time to research and to write this book. They have also politely endured endless scientific and philosophical discussions and patiently allowed themselves to be wired up to computers in hundreds of different cyborg combinations over the past decade.

Richard Soutar
Atlanta, Georgia
2006

Contents

CHAPTER 1	Life Is But A Dream...	1
CHAPTER 2	What is the Self?	7
CHAPTER 3	Habituation	16
CHAPTER 4	Boiling Frogs	22
CHAPTER 5	The Shadow Self	33
CHAPTER 6	Digging Out	47
CHAPTER 7	Waking Up	62
CHAPTER 8	Faces of Confusion	74
CHAPTER 9	The Interactive Self Evaluation Instrument	87
CHAPTER 10	Take The ISI	100

1

Life Is But A Dream…

*Row, Row, Row your Boat
Gently down the stream
Merrily, merrily, merrily, merrily
Life is but a dream*

Many ancient traditions of thought hold that life is a dream or an illusion. Chuang Tsu, a famous Chinese philosopher, wrote of the great Chinese emperor who dreamt he was a butterfly and upon waking said he couldn't be sure whether he was an emperor dreaming he was a butterfly or a butterfly dreaming he was an emperor. Ancient Hindu thought tells us life is an illusion, which they call Maya. Many spin off traditions, such as Buddhism, insist that not only life but also self is an illusion. They argue that the flow of events trick the mind into thinking that a separate self is present and that all suffering is a consequence of this belief in separate self. Christianity tells us not to look for fulfillment in this world but in the kingdom to come. Jesus encouraged us not to worry or lay up treasure in this world, but prepare for an eternal life that was more important, more real. Plato in the Greek tradition said that life was like having your back to a fire in a dark cave and thinking the shadows on the wall were real. Why did the greatest philosophical minds of history see such truth in this position?

Until recently modern science took the exact opposite stance. The scientific materialist tradition asserted that only what could be measured was real. All phenomena had to be explained in terms of cause and effect. Interestingly enough, what was considered measurable a hundred and fifty years ago is very different from today. Our technology has advanced so much that we are aware of and we measure things that were not even thought of when science was young. The idea of brainwaves was highly ridiculed as preposterous when Hans Berger first discovered them in the early twentieth century. Today we topographically map brainwaves for diagnostic purposes without a second thought. In recent times quantum mechanics has forced us to reassess the meaning of cause and effect.

The equations and measurements regarding the difference between waves and particles is challenging our ideas about what is matter and what is energy. The difference between what is solid and real and what is ephemeral appears less significant than what we previously thought. What is real has become once again, as Steven Hawking (1988) notes, a metaphysical question regarding quarks.

Many scientific thinkers today are again finding value in the traditional philosophies of the past. Fred Alan Wolf (1994) has proposed that the universe dreams itself into existence, and supports his theory with some intriguing arguments grounded in the new physics. Michael Talbott (1991) builds on the work of the famous neurosurgeon and researcher Karl Pribram to explore the idea that we live in a holographic reality constructed by the brain itself. Have we come full circle? What are the implications of all this?

When I was young life appeared to be so concrete and real. Now as I look back over fifty years it has a very dreamlike quality. I recall hearing World War II vets who related stories of intense hand-to-hand battles where everything slowed down and became dream-like with a strange attending silence. People often describe intense moments of their life as having a dreamlike quality. Is there something more to this than a trick of the brain? And what does that term a trick of the brain mean? Are brain tricks unreal? Or do they have a basis in neurological process?

The practical side of me says, "What does it matter? I have work to get done in order to survive." But the question plagues us all in quiet and posttraumatic moments, "what's this all about?" "How did I end up here?" Plato argued that the unexamined life was not worth living. So one solution is that I could examine my life minutely with my thought process through philosophical analysis, but that has been done by many other brilliant minds with their own lives over the last several thousand years with no clearly agreed upon conclusions.

Personally, I feel I can achieve a better understanding through scientific analysis, using the techniques and findings of other scientists going back to Galileo. There is less shifting ground and tail chasing in this approach and I feel more secure in the resulting conclusions. After all, the New Philosophy, or Science as it came to be known, was initially a reaction against the endless argument of metaphysics and blind faith of religion. I believe that we are now, in this era, coming into a position to reasonably address these questions in an exploratory manner.

The field of neuropsychology now offers us new possibilities of exploring the nature of the self, the mind, and the meaning of reality. In conjunction with sociology and psychology it offers us a new basis for analysis of human behavior and how to most effectively interpret and direct that behavior for the greatest benefit of all.

In my clinic I have employed the new findings of science with profoundly moving results. What happens in many cases borders on the miraculous. Utilizing EEG biofeedback or neurofeedback, I have seen people who had been given up for lost by other specialists make great strides in recovering their lives. But I also educate them regarding how they work as a person or self, and that too has great transformative power. What many who come for healing don't realize is that they have begun a journey that will eventually confront them with the nature of the self-their self-in a manner that is almost unavoidable. It is this journey into the confrontation with the self that my experience as a clinician has indicated to me is also the pathway to recovery from mental illness. To get better clients must walk this path. Some intuitively recognize it as the encounter they don't want and either stop coming or stop trying, but most are desperate enough to engage it. Those who do are greatly rewarded.

Who Is The Dreamer?

If life *is* a dream, then who is the dreamer? Most people don't ask a lot of questions until things start to fall apart. We start with the American Dream and add our personal dreams and then attempt to create our private dream world. Sometimes we have to build walls around it because others admire our dream world too much and want to take it from us.

As a species we like to daydream a lot. Research tells us that, if we are in the norm, we spend a great deal of time daydreaming, about success and sex or romance-in that order-particularly . We dream about successfully achieving our goals and visions. We may even be doing this while we perform routine job tasks.

Some of us have a habit of dreaming about what we're going to do if things don't work out. We explore all the ways this might happen. We might call this a "day-nightmaring". These internal dramas often have accompanying dialogue that is also very negative. These dark dreams of the day may come to haunt us constantly and wear us down. We ruminate continually about what threatens us. It becomes a habit. We may try to escape through work, or eating, or drinking, or sex, or gambling, but these are only temporary fixes. Why can't we stop our own negative daydreams and self-talk? How can this self of ours be so untogether? Am I falling apart?

We might start losing sleep on a daily basis because our daydreams are so bad we are too upset to sleep. This can be serious. We even may stop dreaming when we do sleep which can impair our learning abilities. If we can't learn, then how can we solve new problems? Many people in this condition just repeat the same behaviors over and over even though they don't work. What is worse, we may even get used to our condition and forget how to be any other way. We may begin flipping in and out of negative daydreams without even noticing we are doing it-and then wonder why we feel so bad. Psychologists call this dissociation. It often leads to anxiety and then depression. Over a third of the population is on this journey. Fifty percent of the population is either heading into or recovering from a mental disorder. Most feel very alone. Even if life is not a dream, we certainly make much of it a dream anyway.

Can The Dreamer Wake Up?

If your dream is bad enough, you often wake from your sleep. How about your daydream? Ancient tradition says that if your dream is difficult enough and you make the effort, you can wake from your dream, the dream you call your life. This is called Liberation, Nirvana, Moksha, Satori, etc. Most traditions, however, insist that in order to do this you must see that the self is an illusion. That you are not who you think you are. It does seem reasonable that a dreamer might be mistaken about his identity if he is in a dream. These traditions also emphasize that this waking up is very difficult to do and requires much effort and discipline. Apparently the thing that is most difficult to overcome is our bad habits of thinking, feeling, and doing. They keep us going in confusing circles. Conflict and friction confront us at every turn. These habits keep leading us back into daydreaming, into escape. This daydream is deeper than we think. We may even call it being awake. We might call it real.

Much of this agrees with modern neuropsychology. My efforts in the following book will be toward bridging the gap between the ancient and modern and discussing how the discoveries of modern science can transform our life and liberate us in the same way as the ancient technologies. But it doesn't have to be such an exclusive enterprise anymore. It does not have to rely on subjective expertise alone. In fact, we have become so dependent on technology that it may be the only way we can do it. Profound transformation and transcendence can now be available to a wide mass of people through modern technology and therefore grounded in public truth. In fact we all seem to be converging on this path more and more. As strange as it sounds our new religion in modern times may be science and it may be that it can include God and the ancient prophets and teachers.

What if you remembered all your dreams at once? In the wink of an eye. Is that possible? I would have never thought it to be so. I never even considered the possibility of such a thing. And yet it happened to me.

I was just putting on my shoes one day when they came to me. I was astounded and confused for months afterwards. I kept wanting to say to myself "this didn't happen" because if it had then I would be forced to fundamentally reconsider my very definition of reality. But it did happen and I can't avoid that experience. Ever since, it seems there is a part of me that is always dreaming. Even now, as I write, here at this desk. Like the famous Emperor I am left asking, "which is more real?"

For one moment I connected with my dreams in a way that forever changed my perspective on life…or I had a small temporal lobe seizure; which can cause such things to be experienced. But what ultimately caused the seizure? It must have a neurophysiological correlate. And why that experience in that way?

If you could wake up what advantage would that have and what would you find? Maybe it is better not to know so much. To roll over and go back to sleep. If you are happy with your life for the moment you might feel "things are going well, why change it?" Or you might feel that everyone else seems to be satisfied with the dream, so who are you to question it? After all, they can't all be wrong.

Many people have periods in their life where things are running along fairly smoothly, but most encounter serious difficulties at least once. In fact the majority of people appear to have crises frequently and then try to forget about it. These moments often start them asking questions about why nice people like themselves should have to suffer these problems. If they suffer serious physical or psychological problems then these questions may plague them daily. Young healthy people who are unattached can often avoid these messy issues, and even the people suffering through them, for quite a long time. Eventually, however, serious problems begin to arise and they are forced to cope with the issues of pain, suffering, and their own mortality. Denial has a limited half-life.

The majority demonstrate their awareness of the issues by turning to religion for answers. The religious solution can be a fairly permanent fix for most people. It offers them a secure and permanent solution and sooths them through sophisticated social, psychological and emotional mechanisms. It provides quick and easy answers to the most disturbing questions and seems to provide a sane and stable template for building and maintaining social order. The majority appear to be satisfied with their "faith" in these received truths. And this is good. Because not everyone is equipped emotionally or psychologically to move beyond this point. However, a fairly continuous trickle of doubters seems to emerge. They are not satisfied with the easy answers and the powerful rituals. Something more seems to be missing.

This book is meant for those who are striking out on their own to discover what they can for themselves. It is meant for those who are looking for profound experiences holding deeper meaning. It is for those who are not afraid to risk the more mundane treasures they could possibly heap up in this world to find and experience something more satisfying, more lasting, and perhaps eternal. It does not suggest turning your back on the world. It does not recommend embracing your "lifeworld" as it is today. It recommends a middle way. It suggests that we have the means to change the suffering in our life into a daily experience of profound insights and personal transformation. It recommends a direct experience of joy that quietly permeates our life in the background, but is usually lost in the turmoil of minds that lack self-regulation. It discusses the ways we become lost, how we do it, and how we might become found. It proposes the possibility of directly experiencing the perfection and unity of life and understanding that experience as a solution. It explores the possibility that personal evolution and growth is never ending. It offers the tacit experience of transcendence. It proposes this can be done through scientific means and that it can be done systematically.

We can learn to spin flax into gold, drink from the Holy Grail, and walk on troubled waters.

2

What is the Self?

Knock, knock.
Whose There?

Many of the great theological and metaphysical traditions focus on the question of self, or Who am I? A famous Zen koan asks, "What did your face look like before you were born?" We tend to point to ourselves and look within for the private source of action we call the self. Jesus said that the kingdom of heaven was within, and then said, seek first the kingdom of heaven and all else shall be delivered unto you. But was he talking about the self, or something that was also within and other than the self? The whole field of psychology pursues this question in its own scientific way and a large part of sociology does so as well.

We have a self and that self often experiences suffering. When people come into my office in crises they begin a journey of self-healing through exploration of the self. It is fundamental to the therapeutic process. So what is this self that is so often the source of so much pain and confusion? And isn't it interesting that psychology and religion both focus on the issue so much? They frequently have very different methods however. I have found that the process of psychological healing is part of a continuum, part of a greater process of self-evolution or maturation. I have found that the journey my clients begin in my office continues on after they leave. They begin that journey with a series of crises that forces them to confront life in a new way, that sends them seeking for help.

Since they cannot continue on in the old way and must seek help, they begin visiting professionals in search of a cure. They may spend months or years going from doctor to doctor, psychologist to psychologist, trying to find an answer to their suffering. In another time they might seek solace from their church, however, this solution does not seem to satisfy their need in present times.

One woman who came to my office in severe depression had even been sent home from the hospital in that state, being told by a psychiatrist (who was a close relative) that there was nothing more they could do for her. When your life is a living hell that is not good news. She was lucky; twelve sessions later her depression was gone. Of course, not everybody heals that quickly. And this has forced me to confront the question, Why not? Another client was a Vietnam vet who was on a half a dozen psychoactive medications for anxiety and depression and lived in a dead end reality dependent on these drugs to get through the day. He took about twenty-five sessions to begin healing. Why longer? I helped a schizophrenic reduce her symptoms to the point where she was fully functional without drugs in ninety sessions but I am still working past that number with an older woman with mild anxiety. I use the same equipment and techniques with all these clients. What makes the difference in recovery time? Why does one self recover faster than another self?

Sometimes it is clearly a question of motivation. We often have husbands drafted by their wives for fixing. Our job is to "straighten him out." One such client smiled nicely each session and reported no change for ninety sessions. Dutifully he proceeded to do very poorly at each trial. He was clearly not very motivated. Later he went to a very good psychotherapist who got in his face and started screaming at him until he became very angry, and according to his therapist wife, he began to heal. Apparently, at some deeper level, even though we think we are motivated, we are not.

What is the nature of this self that has such deep and convoluted levels that it cannot access its true motives? Why does it appear to be at times a divided self? Why do we do things that we keep telling ourselves we don't want to do? Jesus tells us a house divided against itself cannot stand. Certainly many of my clients live in a house very divided against itself. They are full of internal conflict and lost memories. They may have intrusive thoughts that won't go away or even intrusive feelings and voices. Many of our OCD clients have thoughts of doing horrible things to themselves or others, but they can't get them to go away. Some have intrusive alternative personalities that actually take them over. Are these personalities real? A clinic in New York, according to Michael Talbott (1991), brain-mapped individuals undergoing personality shifts and found dramatic changes in their baseline EEGs. Something very neurologically real is going on. It looks like personality is reflected in brainwave patterns. Can the brain develop powerful aggregates or groups of subroutines that constitute partial personalities. Could one group of networks and then another come to dominate brain function like political parties in congress? This might also explain the subject of possession. It suggests the brain develops patterns, at a very high level of order, that it cycles through in some regular fashion in order to sustain its functions.

The power of the brain to direct our behavior beyond our conscious control is impressive. We congratulate ourselves on being directors of our neural fate. We pride ourselves on being in control of our "self." Yet it was clear to Freud as much as a century ago that we often lose control of parts of our "self" as a consequence of emotional trauma. Freud noticed that some patients lost feeling and use of some of their limbs in a manner that defied neurological and anatomical rules regarding nerve pathways. When he hypnotized them, they were able to regain control of those limbs as long as they were not conscious. Antonio Demasio (1999) reports that some of his patients suffering from neurological damage may be buttoning a shirt with one hand as they helplessly watch the other hand unbutton their shirt. This must be the same kind of experience some dieters have as they helplessly watch themselves eat the very cake they know at the moment they don't want to eat.

Christof Koch (2004), in discussing the latest research on consciousness and decision making, notes that there appear to be competing networks in the brain that cycle back and forth until one network wins and expresses the behavior it favors. Our subjective experience is that "I" made a decision. The problem with that experience is that research also shows that our awareness of our decision to do something usually follows a fraction of a second after we begin to initiate an action. More support for the argument that decisions are not consciously made. In fact, this all supports the notion that past decisions regarding thinking, feeling and doing determine our next decision to act based on a nonconscious neural process. This may not eliminate free will, but it certainly reduces it. It changes the whole traditional picture of self and suggests we need to develop different strategies to change human behavior in more positive directions.

We may all be possessed, it just does not appear to be so as long as the brain functions properly. Thomas Saaz (1984) has investigated the evolution of psychiatry and finds a fairly smooth transition between the Inquisition's categorical schemas of possession and early psychiatric diagnostic categories. Many categories of problems were given new and different terminologies, but the recognized cluster of behaviors remained the same. Sociologists make much of this in Labeling Theory, but psychology 101 texts are very open about the parallels between the Inquisition's handbooks and the modern diagnostic manuals. Do our new words explain the process of possession by strange thoughts, feelings, and deeds any better? I don't believe they do until we look at the phenomena from a wider perspective. Addiction may be just a milder form of possession.

I would like to review the concept of self from a multidisciplinary perspective to get a better grasp on the subject matter. In fact my peers have the same feeling.

We are beginning to look at human behavior from the bio-psycho-social perspective in both psychology and sociology. I think it is important to throw in some history and anthropology to round out the investigation as well. As a social psychologist it is my talent to do this artfully. So rather than bury my talents, I will invest them in education.

The Biological Self

Clearly, as the ancient literature suggests, our physical self is the strongest basis we have for concluding that we have a relatively permanent self. We wake up each day with more or less the same face in the mirror. We are separated by skin and space from others. When we physically suffer damage to our body, it is uniquely us that feels the inescapable pain. We recognize or actively anticipate situations that bring pain and avoid them. We seek situations that lead to reward or pleasure. In fact the great philosopher and early scientist Thomas Hobbes recommended a field of psychology based on this observation and John Watson and BF Skinner were to fulfill this vision with the development of behaviorism in early twentieth century psychology.

The early behaviorists dominated the field of psychology because their subject matter of "behavior" was so easy to observe and document. Many of them insisted that all behavior could be explained in terms of the principles of reward and punishment. That left the concept of self exiled into the unsure world of metaphysics. Was the personality, or self, real or just complex responses and reactions to a world of marginally predictable pleasure and pain? Was having a physical self enough? The field of psychology eventually abandoned behavior as the sole dimension, partly as a consequence of research findings but also because it intuitively felt wrong and could not explain the qualitative experience of being a person.

Clearly we have a built in sense of self and other. It appears to have both a biological and social basis. Research has pointed out that the parietal cortex, a section of the brain in the upper back part of the head, is a major processing point in the brain for establishing the ongoing boundary between self and other. Given the complexity of the environment we live in, this is clearly an astounding feat. Andrew Newberg (2001), in his review of the scientific literature, as well as his own research with PET scans, finds this a key area in establishing the physical sense of self. Proprioception, or the feeling of our skin, is not by itself enough. It must be integrated and coordinated with the other senses in order for us to be able to negotiate physical space.

We tend to exist physically in the same context from day to day, living in the same house and going to work in the same office or plant. When we die, our bodies cease their biological process. Our mouths are silenced and we can no longer speak to others. For others our body and speech becomes erased from its context. Is this not what we fear most? To disappear from active participation in the social conversion? This is the point where we begin to enter the lesser know realm of sociological phenomena. The dimension of self begins to overlap the social dimension at this point.

The Social Self

There are huge unknown literatures in sociology on the function of society with relation to the self. Many sociologists have pointed out that society is the vehicle by which our species buffers itself from the environment and organizes its survival capacities. Some sociobiologists have gone so far as to suggest that we are purely vehicles for our genes to maintain their integrity or act out their dance of self-preservation.

As we grow and develop we continually expand our social context on multiple levels. We move from crawling across the living room floor to running around the back yard, then around the neighborhood and soon we are driving around town. Our identity becomes enriched as we absorb more and more of our social world. We further define ourselves through the clothes we wear, the type of work we do, the church we attend, the people we hang out with and the products we buy. We relate to multiple identity groups who interact with us in an ongoing basis. We are a kid, a teen, a parent, a smoker, a poser, a greaser, a preppie, a groupie, etc, etc. We negotiate with others based on our likes and dislikes, our beliefs and values. We grow into our behaviors of thinking, feeling and doing by interacting with our identity group. Through this process we negotiate according to the unwritten rules of norms and mores as well as subtle rules of nonverbal behavior and communication such as body language. We symbolically interact and negotiate our identity all day long through our attire and our gestures. The result is our complex social self.

The Personal (psychological) Self

In developmental psychology this process most clearly begins around the age of two with the word "no." One renowned dramaturgist in sociology has suggested convincingly that "no" may be the most important word in the English language. When we say no, we declare that we are capable of more than just reaction, but can initiate novel behavior. The child becomes aware that it can resist guidance and move in a new direction. It becomes aware on a conscious level that it is a separate self. When this separate self emerges it begins to utilize more complex rules of behavior and consequently develops more complex and subtle emotional vocabularies. It becomes aware that others have a mind and begins to guess at the thoughts of others. The self learns to inhibit strategies for immediate gratification in order to cooperate with the family to survive and achieve higher levels of gratification. The question remains, however, as to how this self works and why it is not always aware of all of its own inner workings.

The Spirit self

Traditional thought frequently refers to an essence within us that in particular identifies us beyond these other dimensions. It is the spirit or soul self. Modern science cannot identify the basis out of which such a matrix or pattern might exist that could contain such a self. How it might sustain its structure outside the body and continue to be dynamic in some dimension is highly problematic based on what we know of the universe at present. However, recently E. Roy John and others have suggested that a quantum ionic plasma field, a sort of energy field, appears to exist around the brain and may be the source of consciousness and memory. This theory, however, still sees consciousness as something that results from the brain process. It is still an epiphenomena of the brain and requires the organic life process for support. In my personal discussions with Karl Pribram (1977) he has confirmed that he also believes in such a field and that it is generated from microtubules (gene movers) in the axons (wiring between the cells) in brain cells. He is presently involved in research with physicists to investigate this process.

Microtubules are a fascinating topic in themselves. They are one of the smallest biological components of the body and appear to be the basic workcrews at the cellular level. When the cells divide they move the genes around during the division process. They can determine how to organize complex proteins faster then the most sophisticated supercomputers. Where do they get this information and how do they store it?

It has recently been proposed by Steven Satinover (2001) that microtubules link quantum events with biological events and that they may allow for an amplification of quantum degrees of freedom. It is noted in physics that most systems at our level of social experience are very mechanical and predetermined in nature. At the quantum level, however, things are not so fixed. The level of freedom for change is much higher. It is a much more flexible dimension of processes. Information may flow from this quantum dimension into ours and back again in a creative process. Since distance is not an issue with regard to communication at this level, there may be connections between events at a gross physical level through this portal. Given the new ideas in String Theory, there may even be communication from other dimensions coming through this quantum membrane. It may even be a channel for consciousness itself, given the right conditions.

Jeffery Schwartz (2002) at UCLA department of psychiatry has worked on a project that suggests that conscious awareness and perhaps intention are not epiphenomena of the brain. His research demonstrates that the brain can intend to change its own structure and function and argues that this indicates that these mechanisms of consciousness and intent must ultimately lie *outside the brain*. He has taught individuals with OCD how to consciously relate to their thinking in a different way and noted changes in brain structure with pre and post MRI analysis. He argues that it is not possible for conscious intention to change the structure of the brain if consciousness itself is a by-product of the brain.

Self-Transformation Versus Self-Transcendence

If waking up takes you beyond yourself into a new state of awareness that answers the question of suffering, then what is the role of the self? If you search deep enough you will find that the answer often comes in the form of a metaphor, hence the emperor's dream, the butterfly's growth and development. The self in this portrait is like a caterpillar that has limited mobility and must push around in the dirt keeping track of all its difficult legs until the day arrives when it begins to spin its cocoon. It then begins a molting period during which a transformation begins to take place. Eventually it breaks free from its cocoon in the form of a beautiful butterfly that can fly freely above the dirt and confusion of the world.

From these more ancient traditions we can see a perspective that portrays the self as a vehicle. This is a very useful concept. It dovetails nicely with research in the behavioral sciences. We clearly cannot operate in the social order without a self. Those who do not have the capacity to negotiate with others according to the appropriate social rules are isolated from participating. They end up in total institutions such as prisons and mental hospitals. This isolation can severely limit their exposure to situations that can enhance the accumulation of growth experiences as well as limit their ability to process experiences in a manner that results in learning.

We are born into a physical and social reality that requires a body and a self to participate. As the body grows and matures, so does the self. We progressively expand the boundaries of our activities and hence our experience as we grow and develop. We learn how to adapt to more varied and complex social environments.

We become progressively expert at solving various personal/social problems unique to the environments we encounter. We master the social realities and mature. We develop in emotional intelligence, wisdom, and compassion. Erik Erickson offered a very sophisticated template as a result of his research to describe these stages of human growth. In the end, he notes, we seek to give back something in return for our experiences; a form of thank you to the world.

As we move along this growth curve we have profound moments of insights. These are often described as epiphanies and transcendent experiences. In these moments we often experience a profound sense of truth and/or beauty. These moments tend to stand out clearly in our memories as defining moments for us. They give our life meaning as they pull the different disparate and jumbled parts of our life experiences into a momentary coherent whole that offers us great clarity. The ancient masters of other traditions tell us this is a taste of something even greater to come if we pursue these insights with dedication and enthusiasm. They describe transcendent states that take us outside and beyond ourselves to something greater and more fulfilling.

For the majority of the few individuals in our culture that actually pursue this direction consciously there is often great frustration. Westerners especially have great difficulty using the more ancient techniques of the east, such as meditation, to achieve these various transcendent states. We tend to jump directly into meditation with great enthusiasm and quickly become impatient with the dubious results. This appears to be because we often skip the prerequisites of personal development and self-regulation required in most eastern traditions. In yoga one is expected to practice physical and moral self-regulation at a high level before engaging in meditative exercises. In the west we want to cut straight to the chase and get the job done. We want to go straight to meditation and high states to escape our problems and we want to do it quickly. Unfortunately it doesn't appear to work that way, unless you take LSD or mescaline; and then you have only a temporary and very distorted fix.

Jack Engler (1986) did some brilliant research and analyses on this issue in the mid 80's. What he found was that westerners tend to get hung up on the personal and cognitive issues when they attempt meditation, while easterners progress very rapidly once they commit to the process. He concluded that eastern cultures prepared individuals better for the meditative process. I have seen the same outcome in my clinic when I attempt to teach people of my culture meditation. We always end up working on personal transformative issues to get them to the point where they can meditate. Others, who come for psychological disorders seem to end up in spiritual pursuits once they resolve the personal issues that lead them into disorder, unless they have severe genetic based disorders.

In the chapters that follow I will explicate these issues in more detail. We will explore all the ways people get lost on their way to social and spiritual maturity. I will do it based on the existing research in the behavioral sciences as well as neuropsychology. Our sciences have evolved to the point where we can put together a pretty comprehensive picture from an interdisciplinary perspective. What emerges is that getting lost is part of the process, its staying lost that is problematical. The only solution to that appears to be suffering.

3

Habituation

Habit is the enormous flywheel of society...
William James

We clearly have a self with multiple dimensions. It is highly complex and has many facets that are reflected in its behavior. It is bordered and bounded at many levels by a world that pushes back upon it and provides feedback. In the past we have been only able to observe and measure the overt behavior of this self. In modern neuropsychology we have been able to move beyond observing behavior alone and can now look inside the head with modern neuroimaging technology such as qEEG, PET scans, and fMRI. We can see the self at work before it manifests as overt behavior. We can watch the brain process information at it manifests thoughts and emotions as well as the intent to speak or move. The line between behavior and interior information processing becomes vague. We can see the self as a form of information processing. Behavior is just one stage in the cycle. Using our knowledge of computers we can begin to model how this information processing systems might work.

It becomes clear that the self uses the body as an information interface or transducer (which is anything that turns energy into physical movement). The body is transparent with respect to information and the self creates opacities, patterns of information rejection and acceptance. These filtering templates define how the self processes information. The body is tuned, through socialization and culture, to allow specific information to pass into the self information system. This self system has specific patterns of activity. The body grows neural structures to maintain and support these patterns. In this manner experience is embedded in the body in the form of percepts-the most basic element of perception-and memories. Robert Ornstein in his classic work on the psychology of meditation elegantly described how the habits of thinking, feeling, and doing form a layer between us and the direct experience of the world. He adapted the concept of "habituation" from psychology to explain this phenomena.

Habit

One of the founders of American psychology, William James, noted that self-consciousness was a like a stream and that the behavior it resulted in was largely habitual in nature. He called habit "the great flywheel of civilization." At a biological level we find it fundamental to the organization of life itself. Enduring, self-repeating patterns of complex chains of molecules begin to become life at a certain level of complexity. To maintain high levels of complexity, every organism must have multiple levels of repetitive patterns. If they are efficient and successful patterns that perform essential tasks related to survival, then the organism endures. It continues its self-replication process and reproduces: it remembers. Replication of an identity pattern is therefore profoundly fundamental to life itself. This is probably why the idea of identity is so deeply embedded in us.

Each new level of complexity in the evolutionary hierarchy of species is likely to offer more feedback. Each new dimension of feedback is likely to generate greater awareness of the environment and consequently the self-awareness of the organism. So, complexity of an organic system leads to higher levels of interaction with the environment, resulting in higher levels of self-awareness or consciousness. Antonio Demasio (1999), who has done a great deal of work in this area, believes this process results in multiple layers of awareness in each organism. The number of layers is dependent upon how evolved the organism is with respect to the phylogenetic scale. In humans he believes that somewhere around the trigeminal plane of the brain (the level of brain stem where the information from facial nerves is received) our consciousness, or self-awareness, achieves the level of intensity that we consider human. It seems to be a matter of thresholds. The question is not which creatures are conscious, but how conscious is each creature? The routines of organic housekeeping at a certain level of complexity of interaction with the environment result inevitably in conscious awareness and self-consciousness.

Demasio notes, surprisingly, that complex human behavior can continue to be performed even if a person is not conscious. Trauma or deficits in certain areas of the brain stem can result in a loss of consciousness. Individuals with certain disorders can perform very complex behaviors while unconscious or asleep. Julian Jaynes (1976) has written about this as well and provides a surprising list of things that can be accomplished without conscious awareness. Demasio concludes that emotion is a distinguishing feature of consciousness.

In fact, this should not really surprise us as most of us frequently catch ourselves daydreaming or thinking about something while we perform a routine task.

If we get too involved with our mental adventures we may make a mistake at our task. Many people report "missing their turn" while driving because they were thinking too hard about something. It is interesting how well many of us can drive a car while having our minds, our attention, completely on something else. Initially it takes all our focus and concentration to even get down the street, but once we learn it becomes automatic, non-self-conscious.

Research on learning indicates we automate tasks once we master them (Schwartz, (2002). When viewing the brain as it learns a new task with Pet scans and the fMRI we can see whole areas of the brain activated (Posner & Raichle, 1994). Once the task is well learned, only small areas activate as the task is performed. The brain is then generating its favorite EEG idling rhythm: alpha. The brain is executing the task automatically while the rest of its architecture rests.

Ask any musician what its like to perform a learned piece of music and they will tell you that if they think about it too much they cannot perform well or even remember the piece. The fingers seem to remember where to go on their own as the musician listens in his mind for the right pattern of tones. We know that a great deal of these finger patterns are stored in the cerebellum of the brain and that special attention networks are activated in the basal ganglia to allow these patterns to emerge as motor activity. If we try to adjust our performance too actively, it usually results in defects. I have seen this happen often to athletes that try to consciously compensate for their possible defects in executional form as they perform. They interrupt the bodies learned execution of the task too much. The result is a broken and interrupted stop and go execution of a task that is defective and inferior.

We forget that we had to learn how to reach or to walk. In fact our memory doesn't even work well until we've developed an awareness of self and begin to talk. Even remembering is a learned process. Recent research indicates we learn how to feel by sitting close to our parents or watching them emote. So we learn to feel in certain patterns as well.

Just as there are rules for playing a piece of music there are rules for performing a certain sequence of behaviors to access environmental resources. To coordinate and orchestrate our whole body in such a complex context requires considerable training and practice. We make many mistakes in acquiring those skills. Much of our initial effort is centered around making and correcting errors. As we become more accurate in our execution of behavioral patterns we require less neural activity to perform it. It becomes routine.

Perception

We tend to organize and routinize behavior in order to get things done quickly and efficiently with as little expenditure of energy as possible. We also tend to do the same thing with perception (Ornstein, 1974). This has subtle and powerful implications that gestalt theorists in psychology have spent over a hundred years studying. Our senses gather raw information, but our brain, below our level of self-awareness, automatically organizes it into meaningful patterns we can consciously use. The most basic of these patterns are part of our biological programming, but in humans they are conditioned to a high degree in the socialization process. Psychologists refer to them as perceptual templates.

We have prototypical patterns of expectation for the shape that dogs and birds and trees will take. We use our parietal lobes to categorize these items. With vision, raw visual information enters our occipital cortex at the back of our head and begins to move forward as it is processed. It is successively filtered through these templates as it moves into higher level conscious awareness in the frontal cortex. The information also moves subcortically through the emotional brain, or limbic system, as it makes its way forward. It all converges in the front where it becomes a full cognitive-emotional picture. Consequently our critically aware self, our executive function, makes conscious decisions regarding the highly processed sensory information with considerable delay, probably at least 300 milliseconds.

While all this processing is going on, a small structure called the amygdala in the emotional brain has already looked at the raw data for patterns that indicate danger and has geared up the entire nervous system accordingly. If a stick looks like a snake, we will have already felt adrenaline and jumped back with racing hearts thanks to this little subcortical guardian. It is only then that the frontal cortex may deliver the information-"not snake." Perhaps this information is too late to prevent you from looking like a scardy-cat, but if it had been real, at least you'd be a live scardy-cat. So, we have a fast and a slow response system. Joseph Le Doux (1996) has investigated this phenomenon quite a bit and we will review it in more detail in a later section.

The perceptual templates (Ornstein,1974) can alter the nature of our experience of object/events considerably. One man's nightmare may be another man's paradise. Culture can teach us to literally see the same object differently, which begs the ontological question, is it the same object? It is obvious to scientists at this point that everyone literally perceives a different world; sometimes radically different. Our perceptual templates help us organize complex input rapidly and efficiently but on the downside may block out a great deal of information that could be useful if our environment has changed suddenly. In such a case we might fail to adapt and, according to Darwin, be deselected in the evolution game.

These templates, then, block out information streams that the brain previously learned to be of little immediate value. And of course, as AntonioDemasio (1994) has pointed out, we need an emotional valuing system even to get a complex organism motivated and moving. The devalued information is discriminated against and ignored. But it may become re-sensitized and reawakened into action if the organism experiences a punishing situation. In learning theory this is covered through the concepts of extinction and spontaneous recovery. In searching for water an animal may go explore certain ponds and settle on the one where it is least threatened, but over time it may encounter other ponds it needs to rely on if the source of threat changes or its main water hole becomes even more dangerous. It is conditioned to stay away from the most dangerous pond, but its fear of other ponds becomes extinguished over time, especially if it revisits them without incident. Classical conditioning, or learning theory, tells us people can learn to become afraid of dogs, or elevators, or cars, even if they need to use them, because of trauma that reprograms the emotional brain.

Even more interesting is that all this takes place below the level of conscious awareness. The part of the brain that is involved in fear conditioning is relatively autonomous from our waking awareness. We can learn to be afraid of things by watching the fear reactions of others without even being aware that the process is taking place. In this manner we can acquire responses that have considerable power over our nervous system that we do not intend to acquire. Situations or environments where we observe these fear conditioning events will then later reproduce the fear in us. In this way we can be conditioned to respond to figures and symbols of authority without knowing why or how. Or we could spontaneously have a panic attack based on a childhood conditioning event and have no conscious awareness of why the panic attack occurred. This phenomenon explains a host of social and psychological processes that we did not fully understand before the 1990's.

The self, then, is a sensory based information system that is selectively absorbing information from the environment. It is doing this in an automatic fashion based on priorities established as a consequence of history of interactions. The prime directive in this system is to avoid pain and access pleasure. There are many rules and exceptions relating to the definitions of pleasure and pain. They clearly vary from context to context. This self system is a recursive process wherein the self-conscious portion of the system decides what percepts it likes and dislikes initially and, by habitually responding emotionally to these percepts, routinizes its responses to the point where they become automatic. At this point they operate below the level of self-awareness. We are confronted with a sensory world we have built up out of sensory selection habits (Ornstein,1974). We feel compelled to respond to this sensory world in a specific set of patterns that habitually get us what we want while avoiding pain and discomfort. The self becomes an automatic response system that operates very fast and mostly below the level of self-awareness.

This conclusion explains why we so often experience so much difficulty trying to change our behavior even when it is in our best interest. It explains why we respond to situations in the same way even after we have determined that we don't want to continue responding that way. It explains why we often become victims of our own unwanted behaviors and addictions. It explains why we become so emotional in the face of circumstances that disrupt our daily routines or threaten our customary ways of thinking, feeling, and doing. It makes evident the reason most of us would rather reject a challenging new piece of information rather than process it and integrate it in a manner that calls for rearranging the whole system. After all, the system is a whole.

4

Boiling Frogs

The truly damned are those who are happy in hell.
George Bernard Shaw

We are putting together an interesting puzzle and I would like to grab for another big piece. The good and bad side of habituation and routinization. The good side of routinization is that it allows us to make sense of and survive a very complex environment. The bad side is that we may over adapt.

The example I like most to use involves boiling frogs. If you drop a frog into a pan of very hot water it will leap out. However, it you put it in warm water and slowly turn up the heat it will continually habituate to the temperature. Folklore has it that you can increase the temperature slowly enough so you can get to the point where the frog will boil and it will fail to detect the danger it is in and fail to escape. Now I don't know if this is true of frogs, I haven't read the literature on this, but I do know it is true of humans in a metaphorical sense.

We know from research that people will adapt to situations that progressively cause more pain up to the point of self-harm. In modern life people adapt to pollution and noise levels that are very dangerous to their health. They also adapt to emotional environments that are destructive to their health. Abused wives get used to the beatings and make excuses for their spouses. Stressful work environments, such as air traffic controllers work in, have been documented to lead to mental disorder and physical tissue damage in the form of ulcers. They may end up taking it out on their wife and kids. J.M. Weiss found that most mammals when put into a laboratory situation where there is a constant no win situation develop internal ulcers that can lead to death. Before that happens Aaron Beck (1979) noted they develop a state of learned helplessness that looks very much like human depression.

Another interesting pitfall of our neurological organization is our relationship to fear. Our brain has a special circuit dedicated to fear. It is located beneath the cortex in an older part of the brain called the limbic system. It has its own memory system and works fairly independently of the conscious mind and the cortex (which may be one and the same). The main player in this system is a small, almond size collection of cells called the amygdala and it has been highly researched up to this point. The existing research indicates it is the center of aggression and defense in mammals.

A few decades ago, a young man climbed to the top of a tower in a college campus in Texas and began shooting people for no apparent reason at all. The story grabbed national media attention and experts from all over began to investigate the case. They could find no hidden reasons for his irrational behavior. He came from a good family, was well socialized, and appeared well adjusted all his life. His girlfriend said he was just becoming more angry and defensive all the time and there seemed to be no explanation. He was aware of the problem and had even gone to a psychiatrist for help. Eventually an autopsy was performed and they discovered a tumor in his amygdala. It was very disturbing to everyone to think that a mere tumor could cause such specific and peculiar behavior. Psychiatrists and neurologists, however, are only too familiar with how damage to specific areas of the brain can cause peculiar and unusual behavior.

The amygdala is forever on guard against danger. It is impulsive and tends to have a stubborn memory. Research (LeDoux, 1996) indicates its attention is strongly attracted to any source of danger and has the ability to manipulate us before we are even aware that something is wrong. It is the topic I bring up when the professional golfers I work with have asked, "Why do my golf balls keep going exactly where I don't want them to go?"

When walking through the woods it was good for early man to constantly be on guard against predators and dangerous situations. The amygdala would remember all the bad stuff and just below our threshold of awareness it would keep focused on any possible source of danger. Its response was quick and automatic. This allowed us to focus on our quarry or other food gathering and exploration tasks at hand. It would, however, take control of our body and force us to jump back from a dangerous snake, even if a few seconds later the conscious cortex discovered it was really just a stick. Mother Nature would rather have us safe than sorry. This circuit of the brain is one of the reasons we have survived as a species, but it is also a frequent source of problems socially. We often overreact to threatening stimuli, such as words and actions of others, before we carefully consider our responses.

Modern man's attention is easily captured, before he is aware of it, by any source of danger. If you are afraid of the rough or the pond, your attention will constantly be drawn to it more than is helpful while you are trying to focus on the green. The result can be a momentary distraction of this system that can throw your aim off and send you golf ball exactly where you don't want it to go. This can be extended to other life situations as well. You may over focus on a threatening looking individual in the street and consequently attract their attention more than you want to and they may be drawn to your restless and fearful body language.

What makes matters worse for humans is that the amygdala has trouble discerning between internal and external sources of fear. Some of us get so distracted and preoccupied by the internal sources of fear that we have difficulty concentrating on daily activities. These sources come to us in the form of images and internal dialogue and they are often intrusive and uninvited as far as we can tell. What we don't realize is that the amygdala is just doing what our parents told it to do when they had more control of our brains than we did. This is not to say that they necessarily did this consciously and on purpose. We learned by watching them and they were often at the mercy of their own amygdalas. The interesting aspect of this is that this process represents an unconscious intergenerational method for transferring information regarding danger. It is an automatic system that operates with considerable independence from our ongoing conscious considerations and control. Our fearful behavior is taught and learned without any conscious effort. It is automatic.

An interesting story exists in the literature that demonstrates just how independent and unconscious the amygdalic network happens to be (Changeux, 1985). It involves a famous neurosurgeon who had a patient that had experienced severe damage to their conscious memory system and they could not make any new memories. The patient could however remember everything about the past that she had consolidated prior to the damage to her brain. Each day this surgeon had to introduce himself for the first time and the patient likewise responded as if she had never met him before. This surgeon, however, had noted that the patient did seem to have some kind of memory for aversive stimuli. So, he put a sharp object in his hand one day and shook her hand during the introduction. The pain forced her to recoil in anger. The following day he tried the same thing. Again, she behaved as if she had never met him, but when he put his hand out she refused to shake his hand even though she could not explain why.

Classical Conditioning and the Amygdala

Joseph LeDoux (2002) has done a great deal of research with this topic and has concluded that the amygdala is largely programmed by a type of learning that we call classical conditioning in psychology. Once we make a strong association with an attractive or aversive stimulus, it will inspire the same response in us every time it appears. If we are bitten by a black dog when we are a child we will respond to all black dogs in the future as if they are going to bite us. We learn a response that contributes to our survival and avoidance of pain and we generalize it.

The amygdalic system is not rigidly confined to past patterns of behavior, however, and can adapt to new and complex situations; but this change is often difficult to install. It may require a great deal of exposure to the perceived source of danger as well as pleasurable experience relating to it before a change occurs. We may end up raising our own black puppy into a dog and learn not to respond fearfully to our own black dog. Other black dogs, however, may continue to inspire considerable discomfort in us.

Humans of course learn a great deal through a process which Albert Bandura called vicarious learning. We do not have to be bitten by a black dog to become afraid of them. If we see our mother responding fearfully to black dogs when we are growing up, we will be conditioned to fear them as well. The fears of the parents are passed on to the children through vicarious classical conditioning. By watching others we learn vicariously what to fear and this often happens before we are old enough to even be consciously aware of the event taking place. We are programmed, in a sense, to respond automatically as little children with very little say about it all. If we do not review this automatic behavior in a conscious and critical manner, it continues to control us automatically.

A One Way Street

Le Doux notes that there are more pathways running from the amygdala to the cortex than vice versa. He interprets this to mean that the amygdala can easily gain the upper hand over the cortex. When it comes to fear, the cortex is at a disadvantage. The greater the fear, the more the disadvantage. In fact, in moments of intense fear we may engage in behaviors that we would never engage in if we "were in our right minds," as we like to say. One of my psychology texts cites the story of a man who accidentally shot his daughter in a moment of fearful doubt, thinking she was a burglar. The amygdala pulled the trigger in this case, but without him knowing this-imagine the guilt he experienced. Consider how this guilt defined his behavior and his beliefs about himself. We often say and do things in anger and fear that we later regret. This network in the brain explains how that can be. We are forced to conclude that habituation and routinization, when taken too far, can lead to serious and unnecessary conflict and death.

Fear and Stress

Fear, as we can see, is automatic. Once our circuits are programmed, it is fast and mechanical in its response. It also has serious ramifications with respect to our body, since it is the body that must expend the energy to react in response to a dangerous stimulus. The amygdala has direct connections to a part of the nervous system known as the autonomic nervous system. This system has two parts. One part, called the sympathetic system, gets the body ready for fight or flight. The other part, the parasympathetic system, calms the body down. As we go through our day, we are constantly shifting from states of arousal to states of relaxation. This constant shifting is in response to both internal and external events that excite the amygdala. It apparently cannot distinguish the difference between inner and outer. That job is for the cortex to do but that job does not get done until after the whole system has been mobilized for fight or flight.

When the body gears up for danger many things happen. Blood sugar levels increase as the liver prepares the body for heavy activity. Muscles tense up. We breathe faster and our heart speeds up. We sweat and our mouth becomes dry. Peripheral blood vessels constrict to protect us from excess bleeding in case we are physically harmed. The immune system functions are temporarily suppressed so all energy can be focused on immediate survival needs. The adrenal glands release a great deal of norepinephrine and cortisol. You might notice many of these changes occurring whenever you go to speak or perform in front of an audience. This is a very taxing state for the body and it is not designed to sustain this metabolic effort for very long. If it goes on too long the body begins to use up all of its reserves and weaken.

The brain responds during the initial arousal phase by reducing conscious executive functions in the prefrontal region that might slow down automatic reflexes. In this situation the individual well trained in efficient automatic responses to danger will do best. We understand this well as a species. We train individuals in combat, martial arts, and medical emergency procedures. This allows individuals to function well under great stress when their executive functions are hampered by over arousal. Interestingly enough, we don't train individuals for interpersonal conflict resolution, especially in intimate relations, yet this is one of the key areas of human conflict. A 50% divorce rate tells us so.

Social Distress

For animals, the sources of stress are fairly limited and routine. Animals usually have to find food and fight off occasional predators. They have predictable routines and their adversaries are known. For humans the sources are very complex and unpredictable. We have highly varied schedules that are intensely demanding and contain constant unknowns. As social animals, we have developed very complex interactions that take place on both a physical and a symbolic level. Each of those levels has multiple dimensions of its own because the two constantly interact. We are focused on both an inner drama and an outer drama. The two may be taking place simultaneously or one may be temporarily dominating the other. Either way, our older brain structures and our body cannot tell the difference. Your limbic system and your body respond the same way to a bounced check as it does to a bear in the woods.

Research suggests that we are constantly creating our social reality in an ongoing manner. A rock may be temporarily defined as a table, a chair, a podium, a pulpit, or an impediment, to name just a few. Depending on the actor and his claims, the object can shift in meaning from moment to moment depending on who is dominating the social conversation. If any of these meanings are a threat to us, the amygdala will respond with heightened arousal. The density of our modern daily round of social exchange is such that we encounter a great deal of disturbing stimuli in the course of a day. This means a lot of upsets and if we are not able to effectively deal with them the consequences are "the cycle of failure" and "learned helplessness."

The Cycle of Failure

This is a term I have come up with to describe a process that takes place when we fail to successfully solve a problem in an expected time frame. Notice again the word "expectation" appears. We define all situations with expectation so we can prepare different courses or lines of action. We automatically prepare to some degree, if we can, based on past experience. We dislike the unknown because it can leave us unprepared and with minimal control over a threatening situation. If a situation presents the possibility of familiar negotiations we often quickly review the possibilities and our set response patterns. We ready our tools for the job, if you will. If the situation is unfamiliar then we will begin reviewing possible courses of action and start some preliminary problem solving.

Now several things can go wrong at this point. If new situations have always resulted in failure, humiliation, or physical abuse, then our fear system will already be up and running and we will be hyperaroused. If in those previous situations all of our alternative lines of action were met with failure, then we will begin a full scale assault on problem solving. Our minds will go round and round in a wild effort to come to a solution. Usually we are so aroused at this point that it interferes with our actual ability to problem solve. In addition, to respond to a novel solution requires considerable skill without working it out first at a physical level through trial and error. We tend to rely on a fixed set of solutions in an automatic manner. So we keep frantically rummaging through our tool box picking up tool after tool and saying to ourselves "nope" "nope." In psychology we call this "perseverance" and "having fixed sets". As we review our options, our fear and arousal increases. We have a feeling that we are headed over the cliff and we can't find the breaks. The cortex works less and less effectively as the amygdala pressures it to come to a solution and act quickly. The body and the emotional brain think you are about to die, when all you may be doing is trying to think of the right thing to do in an awkward social situation.

If you encounter failure often enough and the dire consequences that follow, you may fly into a full panic every time you encounter the stimulus or situation. The definition of the situation may be that it is time for fight or flight. Of course most social situations make that difficult without some prior negotiation. So advanced avoidance of any threatening situation is often the preferred course of action.

These internal responses lead to very negative self-evaluations. The expected dire consequences are avoided, but resulting shame and guilt cannot be avoided. The result is a situation wherein you lose either way. We call this a double-bind. In this kind of situation most mammals become aggressive. After a time they become exhausted from the stress, give up and lie down passively until they die. As far as we know most mammals don't feel guilt and shame and they do not learn vicariously or have complex internal shifting definitions of the situation. They do however respond to no win situations the same as humans.

Fear Sets

What are the set of items that compose your personal constellation of fears? Over time, as we are socialized, we acquire our own dictionary of fear. We learn to be afraid of types of people, things, and situations because they present to us stimuli that have been classically conditioned to elicit fear in us. It is not a rational process. We construct our daily lives to avoid these fear sets. We build multiple levels of routines that buffer us as much as possible from these fear sets. Each time a level of our routine is breached we become more fearful. We construct comfort zones of activity and develop strategies to maintain their boundaries. This is all done over many years of development and it is only a partially conscious process. Most of the behavior is executed with little or no self-awareness. It is automatic.

Self-Fulfilling Prophecy

We have found from social research that teachers who expect students to do poorly end up setting those students up for failure without being aware that this process is taking place. This is one more demonstration of our automatic nature. Because of our automatic nature, our own strategies get turned against us within our internal dialogue. If we believe no one can be trusted, we tend to act in a reserved and distrustful manner that often puts others on their guard. They tend to be wary of us and we become distrusted. Their growing attitude over time tends to prove to us that they are truly not worthy of trust. We look for evidence to support our hypothesis without even realizing it. This is called "belief perseverance" in psychology.

The interesting thing about this process is that the face we wear that conveys mistrust is learned and automatically becomes part of our repertoire. It appears automatically in open conversation, as do most facial expressions to facilitate communication. We may hide it, but frequently we let our guard down or give ourselves away through some other gesture. If we practice hiding all our emotions, as some people who are raised in emotionally abusive families do, then others will also mistrust us because they cannot read us well.

It has become clear from this that whatever we are capable of doing to others we end up doing to ourselves. We apparently live in a social order that is a feedback system. If we are raised in an environment where trust is low, then we tend to mistrust others. Others become aware of this through body language, paralanguage, and bioelectric fields that interface as we approach others, such as the electromagnetic pulse field generated by the heart. This is a multi-level interface that is both conscious and non-conscious in its response. Others then respond with mistrust toward us and so we become the mistrusted in that system of social interaction. How we treat others becomes how we are treated. Even more fundamental is that our beliefs drive our behavior. So what we believe comes true.

This brings up another interesting point. If we trust others, then others will tend to trust us. If we respect others, then others will tend to respect us. We cannot get trust through surveillance or respect through coercion. We get the reverse. The question then arises, "Once in a negative feedback loop, how does one get free?" It appears to take a crisis that leads to insight and a leap of faith. We must then fake it until we make it. It takes faith in information coming from outside our socialization process or someone who demonstrates unconditional trust and respect. This explains a lot about the traditional wisdom embedded in religious traditions.

Core Beliefs

I bring up the issue of mistrust because I find it a common problem for people with disorders. In couples counseling it is often the key issue blocking the relationship. Mistrust leads to frustration and anger. Mistrust grows out of fear. The fear is usually one of emotional pain. Couples stop talking to each other because they develop so many rules about what the other cannot say without fear of reprisal. When communications dry up we start interacting with the other person in our head and their dialogue is patterned after our belief that they will lead us into conflict. A double-bind once again.

Aaron Beck's (1979) research has lead many of us who deal in both research and counseling to the conclusion that much of the negative thoughts and emotions individuals have arise out of habit. The habits are supported by beliefs or assumptions. For Beck these assumptions are usually some version of "I am helpless" or "I am unlovable." I too find this true in my practice. One would think that just pointing this out to others would solve the problem, but it does not. To say to an individual "stop thinking that way is the same as saying "stop feeling that way." For humans it cannot be done. It is a routine that is embedded in their neurological structure and it must be grown out of and not just stopped. It must be learned.

On average we find it takes at least a month to initially break a habit, to grow out of it. Research in learning shows us extinction, the ending of a routine behavior, occurs only slowly and only if reinforcement of that behavior is not there. To eliminate the reinforcement in humans we must alter the internal stimuli as well. First an individual must become aware of a specific pattern of internal dialogue such as "I knew it, things never turn out right for me." Then one must learn a new skill; reviewing the validity of the automatic thought and challenging it. Then a substitute automatic thought, such as "I got it wrong this time but now I'll know how to do it better next time," must be established through constant practice. This requires a great deal of effort to do. Most people require a lot of motivation and support to accomplish this as well as the guidance of a professional.

Brainwave Training

An interesting solution to this problem has emerged recently in the neurosciences; it is called "brainwave training." What we have found with brainwave training, or Neurofeedback, is that individuals become very self-aware as a consequence of the training and start catching themselves in their automatic thoughts.

They are usually tipped off because they become aware of the attendant negative feelings. Rather than continue in the pattern, they drop it spontaneously. This is because the brainwave training they have had teaches them to avoid dissociating or perseverating. Their brain prefers a neutral state of rest. It naturally returns to this state after each task. Individuals who worry, ruminate, or dissociate have difficulty maintaining this resting state. Their brain forgets how to get into this state and they get stuck in other negative and energy draining states such as constant worry. Brainwave training teaches them how to find this resting state again and get in the habit of returning to it when the task at hand is done. It is an attractor state that has a specific brainwave signature and once its has been re-learned and re-established, it tends to dominate a brain at rest.

Neurofeedback is technically EEG biofeedback or brainwave biofeedback. Each brain state we have has a specific signature in terms of electrical activity that shows up in brainwaves. Presently we know what the more or less normal patterns look like as well as what the abnormal patterns look like. Using computers to monitor these different signatures or brainwave patterns, we can teach individuals to change them at will and gain greater control over their patterns. This gives them an enhanced ability to manage their moods, thoughts, attention and other functions. In fact, many professional and Olympic athletes are now using this technology to enhance their performance. Even high level business executives are taking advantage of it. You may wonder why you haven't heard more about it. If you had such a technology that gave you such a competitive edge, who would you tell about it? Fortunately this technology is being used very successfully to remediate Attention Deficit Disorder, Anxiety, Depression and a host of other problems. It is slowly gaining a reputation in this area and people are beginning to become aware of it. We will discuss this topic in more detail in a later chapter.

Social Context

Another impediment to growth out of old patterns is the social environment the individual is living in. Most individuals with serious problems have constructed a social context that reflects and supports their issues. They have not constructed it consciously but as a consequence of the thoughts, feelings, and actions that have issued from their perspective and underlying beliefs. It is a social reflection of their interior life. Self and social context are two interdependent dimensions or systems that dovetail precisely. They are part of the bio-psycho-social matrix of nested systems. Self and social context are in a constant feedback process. Each system attempts to maintain its own structure and integrity.

People often think of the social context as a passive process, but it is supported by other selves in constant negotiation at all levels and therefore is somewhat externally coercive. If we have a spouse or roommate, we are constantly negotiating at multiple levels regarding the environment. Who takes out the garbage, how is the toothpaste tube used, who rinses the dishes, what brand of toilet paper is used, who gets to watch TV and when, how much emotion is safe to show, what subjects should not be talked about, how much attention each person requires, and even far more subtle aspects of body language and paralanguage can be used as well as semantic structures of verbal exchange. This rich and invisible context can provide counter-training in an ongoing manner because it will renegotiate the client back into their old behavior patterns. Consequently this context must be identified and strategically altered in key aspects through efforts of both trainer or coach and client if sustained change is to occur. Friends and even lovers may be lost, jobs given up, and relocation negotiated. All of these must be considered in the light of supporting new behavior and extinction of old behaviors.

It is this disregarded dimension in psycho-therapy, and even weight loss programs, that often confounds efforts to change. In our culture we like to think of ourselves as relatively independent agents who are free to change at any time. However, the social research empirically demonstrates that we continually negotiate our identity with others through a medium of physical, social, and mental objects. Their impact on our behavior is profound. If you go to the doctor for indigestion but fail to change your diet and eating habits, then there is nothing the doctor can do in the long run other than prescribe medication to cope with it.

In Summary

We have covered a variety of ideas here, but they all relate to the topic of how stress enters our lives as human beings. It impacts us at many levels and in many dimensions because we are such complex beings. Our physiological response to stress translates through a variety of mechanisms, mostly automatic and manifest as habit and routine, into the realms of the psychological and the resultant social interactions. This translation has key features that often confound our lives. Fear and how we deal with it is a key issue. At the core of this process is "trust" and the complementary drive for control, as the celebrated developmental psychologist Erik Erikson (1968) was so well aware of when he placed it at the beginning of his progression of dimensions of human development throughout the life span. I have followed a thread here, that although at first may appear thin, will grow fatter and stronger as we continue to develop our theme. It is a difficult maze to navigate, as one famous Greek hero discovered.

5

The Shadow Self

Knavery's plain face is never seen 'til used.
Shakespeare

The consequences of our neurological structure are far reaching. It impacts every facet of our life. We can see this structure reflected in everything we do because it is the source of our behavior. When problems arise in our life, knowledge of it can provide insights into potential solutions that can be very effective. I have used this knowledge extensively in helping others recover from mental disorder.

Socialization

How we are socialized influences the structure of our brain. Research suggests that the first two years of life are crucial for growing the frontal lobe circuitry of the brain for managing emotions (Sshore, 1994). Lack of appropriate emotional stimulation can result in difficulty with emotional development. Children neglected during this period often become depressed adults.

This is also the period when we begin to learn the emotional scripts of our parents. We watch how they use their emotions and learn to imitate their patterns. In fact infants and young children will mimic the arousal patterns of the physically closest parent. I have often demonstrated this by hooking a parent and child up to individual SCR channels and measuring their emotional arousal levels as they interact. SCR is a measurement of what has been called the galvanic skin response: a shift in sweat gland activity due to stress or excitement that changes the resistance of the skin and that can be used as a measure of arousal. Antonio Demasio (1994) has used this as a measure of emotional responsivity in his own research. As we become emotionally aroused, the information is apparently passed on to the cortex through the ventromedial prefrontal cortex, which is in the front middle of the brain, into our conscious awareness. During this process the information is also passed on to the hypothalamus and then to the endocrine system which increases the sweat gland activity that we measure as SCR.

Initially, during an interaction, there is no great similarity in their patterns of arousal and this can be seen on a computer screen in the form of two wavy lines criss-crossing each other at random. As parent and child interact more and more you can see the arousal patterns moving together until moving lines overlap each other in the exact same pattern. This means their physiological patterns in their autonomic systems, a part of their central nervous system, are working in synchrony together as if they were the same body. My biofeedback mentor, Marjorie Toomim at the Biofeedback Institute of Los Angeles, showed me this amazing phenomenon.

When I was just finishing up my dissertation in graduate school I had become, like many other graduate students, very distressed because of all the pressures involved. My daughter and I were standing next to each other doing the dishes when she said that she was having trouble breathing. I asked her to go across the kitchen and into the living room for a few minutes and asked her to tell me how she felt. She responded that she felt fine and could breath again. I then asked her to return to the kitchen to finish her job. Again she reported the difficulty breathing. Her nervous system was resonating with mine and tracking it automatically below our levels of self awareness. Think about how many times you may have been upset as a child and not sure why. If you are upset and don't have an immediate reason you may find yourself inventing one. You might make up a story in your mind to explain your feelings that is erroneous. At the very least, the arousal will call forth a related emotional situation from your memory which will engage you emotionally further.

As children, our training takes place at the emotional level by engaging our parents and siblings. The emotive patterns of arousal become conditioned to specific contexts, they are routinized and become automatic. Before the self-awareness system in the brain even comes into existence around the age of two, the patterns have already begun to be established. As verbal skills expand during the fast mapping period around ages four and five, the meaning of these emotional contexts is expanded and further anchored in the subcortical processing system.

What the child sees and experiences is given differential emotional value by the parent. The child learns vicariously by watching what to fear and what to seek for pleasure. After the age of two, the basic emotions (ie fear, disgust, joy, sadness, and anger) which are hardwired into the brain, become combined into more complex emotions that require greater self-awareness. It is very difficult, for instance, to have shame unless one has a self to be ashamed of. These more complex emotions soon become routinized as well.

The Faster Gun

Since the amygdala has its own memory system and responds to stimuli faster than the cortex, some very interesting consequences ensue when we consider human interaction. It has the ability to interpret the behavior of others below the threshold of conscious awareness and to respond very quickly without permission from the cortex. In the world of complex human exchange, that could result in some very erroneous responses. The cortex is then left with the job of mopping up the consequences and the results are often unpleasant.

The first area of interest with respect to this feature of behavior is language. Gestures are an important part of language and many gestures are routine behaviors emitted with little awareness involved. Many experts in the field of communications believe that as much as *ninety percent* of communication is non-verbal. This suggests that most communication takes place automatically and without our awareness. If this is the case, then we are often at the mercy of our habitual responses during most exchanges. We are reduced almost to the point of helpless observers at the time. A very fast game is taking place and we have a limited view of it and a delayed input. To make matters worse, that input is diminished as the stakes in the outcome increase with escalating arousal.

Most animals have their own meaningful gestures, but humans take it further and incorporate it into verbal symbols. It involves tone or prosody, tempo, accent, etc. This aspect of language is often called paralanguage. By stressing different tones and accents with the sentence "How are you today?" You can generate meanings that vary from cold cordiality, to intimate invitation to rude sarcasm. We do this casually and without much attention to the details. This aspect of language provides us with a greater density of information but also provides another arena for automatic behavior and the confusion that often results.

As children, we encode in our memory systems the phonetic and tonal patterns our parents and siblings use without discrimination. We tend to use them when the situation elicits them from us. The meaning of these patterns may vary greatly from family to family and culture to culture. They become automatic segments of behavior that we use to respond. One family's neutral response, however, may be another family's threat. When others use similar words, tones, or facial expressions, our interpretation is rapid and automatic. The amygdala may pick up on certain features of another person's communication and become threatened. Before the other person has even had a chance to finish their communication we may be already responding. Our response very likely occurs before we have entirely evaluated their response with full cognitive awareness processed in the cortex.

At this point, the other person's amygdala may have processed the threatening response and sent forth it's own volley of defensive behavior. At the same time it is increasing the arousal level of the central nervous system, preparing for fight or flight. The initial response of the cortex to the amygdala's mobilization is to speed up the cognitive process to accelerate problem solving. If the arousal is too intense then the cortex will be overwhelmed and automatic systems will increasingly take over the responses. As arousal increases in both parties, the cortex becomes increasingly hampered and a full blown battle will ensue until one or both parties can gain control over the situation or their emotional system.

As can be seen, much of what constitutes arguments is automatic ritual behavior. The manner in which the response components are strung together is often grounded in familial patterns as well. The typical argument between two individuals becomes a stereotypical series of responses much like a dance in which the performers have little control and during which they may even be watching helplessly to some degree. Once the sequence is done, they may calm down and gain control again: often with attendant feelings of regret, frustration, and guilt.

This explains why it is so difficult to work with couples in counseling to get them to change their behavior. They lack awareness of the details of their responses and have little control over them. This suggests that one of the best methods to resolve this problem may be self-awareness training in conjunction with repetitious training in new responses. Practicing a new response set so that when aroused they have alternative patterns to draw upon would be a crucial feature of changing behavior. This is much like what takes place in sports training. Arousal levels and response patterns are practiced over and over again for the coming event. Unfortunately, even then old behaviors may intrude if arousal levels become too high.

Deferring Gratification

Learning to inhibit and regulate the emotional brain is a key feature of maturity. It is a very important aspect of socialization according to the sociologist Travis Hirschi (1992). He and his co-author, a psychologist by the name of Gottfredson, believe that deviant behavior often arises from this inability to defer gratification. The child must have it modeled by the parents and then learn it themselves through parental intervention. They learn that such deferred gratification benefits both them and the family.

Unfortunately, if one or both parents have difficulty inhibiting their own impulses, then the child may learn the exact opposite. This inhibition problem may exist in the adult due to excess arousal from distress, chronic pain related to disease or injury, excess drug or alcohol use, PTSD, or head injury. All these phenomena may generate increased slow wave activity in the frontal cortex, the area that regulates emotional behavior. This loss of cortical inhibitory power reduces control over somatic and limbic centers.

Often these causes of inhibitory control problems come together synergistically. A child may be physically abused and receive repeated head injuries that do not necessarily have to cause unconsciousness to generate significant damage. The emotional trauma can also cause post-traumatic stress disorder or physical changes in the brain leading to altered function (Shore, 1994). The aberrant social process thus becomes reflected and physically encoded in the brain. As the child grows older it may seek extra stimulation to compensate for these deficits or engage impulsively in dangerous or pleasure seeking behaviors. Drugs and alcohol also become attractive alternatives to those suffering anxiety and depression. They frequently change brain function enough initially to make the child feel more normal. Amphetamines improve the function of those with AD/HD type symptoms while the initial dopamine elevations sooth those with anxiety issues or Reward Deficiency Syndrome (RDS). RDS is a condition described by biochemist Ken Blum (1996) in which individuals do not have enough of the neurotransmitter dopamine in their system and as a consequence are not able to experience the intrinsic reward that people find in everyday activities. This condition is a result of a genetic variant. Dopamine is the neurotransmitter that increases when we feel good about what we are doing. When you get excited and enjoy talking to your friends, it is a consequence of increases in dopamine in the pleasure centers in your brain. Some individuals do not find pleasure in any routine social behavior and consequently seek extreme conditions or situations in order to feel positively aroused. This can lead to fairly deviant behavior and identities.

Without the ability to inhibit impulses and defer gratification that often accompanies these neurological deficits, a child may fail at paying attention in class, following rules, co-operative interaction with peers, and on task performance. This in turn generates more anxiety and frustration in the child that often results in even greater levels of anger and impulsivity. Adults rarely understand that this is a neurological condition and personalize their interactions with children of this type. They may angrily demand "When are you going to start paying attention? Why won't you do what your told? How many times do I have to…?" and so forth and so on. This again becomes further abuse and punishment to the child. If the parent has similar neurological deficits the interaction may frequently escalate into physical abuse.

These patterns of impulsive response are usually those patterns that have emerged through operant conditioning in the child's environment of socialization. The child learns through watching and or interacting how to respond in a manner that communicates in a visceral manner what the child feels or thinks it needs. Very often what the child really needs, according to Dinkmeyer and McKay (1976), is positive attention, loving contact, or safety. These are often the core needs of adults as well. The combinations and permutations of social situations that can be used to play out this simple drama, as any good playwright knows, are infinite. It is very confusing for most of us to begin to sort it out. However, if you know the basic pattern you are looking for, it becomes much easier.

For most human beings, if we cannot get the positive aspects of these needs met, we will usually settle for the negative aspects. We develop sophisticated strategies, usually borrowed from watching others, to intuitively achieve our goals. These more complex strategies involve a series of manipulative steps. Each step of our strategy then becomes our goal, rather than the end goal itself. As the sequence becomes progressively rehearsed and automatic we become less aware of our implementation of the strategy as well as the process of execution. We confuse the means with the end and often forget why we developed the means to begin with. An evasive answer, a provocative remark, a sarcastic reply; all become tools to achieve forgotten ends. For example a husband might comment to his wife that the porch looks very disorganized in order to manipulate her into cleaning it rather than say how uncomfortable he feels when the house is not clean and ask her if he can assist her in cleaning it.

There appears to be two types of manipulations beside physical threat; rational and emotional. The rational manipulators will propose a course of action and patiently and consistently provide counter arguments and biased information until they wear down the other person, who they always see in an adversarial light, and get them to agree. The emotional manipulators will portray a situation using strong language, emotions and images to instill a feeling of frustration so that they inspire others to make the same decisions as they themselves have arrived at regarding the situation. Individuals using these types of approaches often don't know how they really feel about each issue in a situation but instead see every aspect of the situation they think about as critical. This is because they thought of it or felt it and it must be right, since to be wrong would be to have a defect or be in danger.

Much of my work with adults from traumatic backgrounds involves identifying these strategies and rediscovering their underlying goal. Once the adult networks of our brain see the ineffectiveness of these child based strategies it is easier for them to inhibit them and implement new strategies. It is the job of the therapist to suggest new strategies that are more direct and grounded in mutual trust and compromise.

Strategies That Don't Work

In dysfunctional and non-optimal environments of socialization communication is often minimal or ineffective. Researchers in family structure and function find that communication is a key process facilitating cohesion and adaptation. When communication is good, then each person is aware of the other's likes and dislikes and of the best way to negotiate getting their needs met. Forbearance, personal sacrifice, compassion, consideration of others, etiquette, and delayed gratification are part of the negotiation process. In extreme environments this can be transformed into becoming a doormat or a tyrant. The extreme conditions tend to lead to mistrust and a context wherein individuals hide their motives and their feelings. This secrecy leads to reduced interaction and very strategic interaction, as described above, aimed at manipulation and control. From this environment specialized strategies emerge unique to the situation. Children socialized with these rigid strategic strategies often discover these strategies do not work outside their family system. They offend others and generate mistrust. This however confirms their family assumption that others are not to be trusted.

Over time, as mentioned previously, individuals in this situation become habituated to the learned patterns of interaction and focus on the execution of their strategy to get what they want, rather than focus on their original purpose and goal (ie safety, love, attention, etc.).

Family Types

Rigidly disengaged or chaotically enmeshed families generate rigid roles and scripts. Flexible families allow everyone to try out and exchange roles continually. Children who observe the dysfunctional versions in their socialization process do not get to try out alternative roles that lead to successful interaction and assume that many are not available to them. This limits their skill at human interaction within their society and reduces their chances of successful interaction and intimacy. Skills at personal self-disclosure which lead to friendship and intimacy may either be absent or sabotaged by their own automatic habitual response patterns to what they perceive as threatening cues.

In rigidly disengaged families the roles are fixed and unchanging. Interaction is limited and fairly unemotional. Emotions are hidden and devalued and often considered a sign of immaturity or weakness. Showing emotions may lead to criticism or avoidance on the part of others. Displayed emotions observed by others may be used by them as a source of information and means to manipulate the individual who displayed them at a later time.

In chaotically enmeshed families everyone is feeling everyone else's emotions constantly and it is difficult to determine who is the source. Roles become interchanged inappropriately and constantly. Nobody knows who they will be called on to be from one moment to the next. Individuals are constantly intruding into each other's space. Privacy is almost non-existent. Boundaries are constantly violated. Everyone is expected to know everyone else's needs and feelings and act in accordance.

These types of extreme socializing environments result in automatic habits of thinking, feeling, and doing that inevitably lead to conflict and failed relationships outside the family of origin. Individuals engaging in the behaviors learned from these environments usually have no idea they are continuing to play out the family scripts and are oblivious to the patterns behind them.

Telling Others Who They Are

In the process of manipulating others to achieve each step in their automatic strategies, individuals from extreme family types will focus a great deal on defining others in negative terms in the course of their daily interactions. They often use terms such as "always," "never," and "every time." These terms reflect a form of black and white thinking as well as a cortex that is very rigid and threatened by emotional arousal. Research shows we develop and use the right front of our brain to control our emotions (Shore, 1994). Individuals with frontal head injuries or excessive exposure to trauma as children may have problems with this region of their brain. With limited inhibitory ability in the right hemisphere, this type of cortex is easily overwhelmed by emotion and finds it difficult to calm down once emotion has emerged in any robust form. There is little ground for compromise with black and white thinking and individuals who engage in it are extremely threatened by disagreement. If others choose not to favor the option they have embraced, it suggests that they themselves are defective because they have made an erroneous decision. It also suggests that the other person may reject them because there is disagreement. It further indicates that the other person may not play along with them and is not to be trusted. Finally, they may be at risk because they have made an error and errors usually lead to very dangerous situations and lack of control. For these individuals communication is a minefield and absolute loyalty is necessary to proceed. Rather than speak in terms of "I" and reveal their own feelings in pursuit of compromise, their mistrust drives them to manipulate the identity of others. If they have PTSD, they will often shift quickly into rage to protect themselves. Statements usually begin with "you" and take the form of "You never listen to me," "Your always late," "You don't care about me, you only care about yourself." These statements are unconsciously designed to hurt and provoke others into response patterns that can be further manipulated. They authoritatively decree a definition for the other person that feels absolute and final. This is, in a sense, "telling the person who they are." This inevitably elicits a defensive response from that person as they attempt to explain their behavior and cast themselves in a more reasonable light. Often the response is one of frustration and anger at having been portrayed so negatively and not being worthy of the mutual trust a relationship implies. This type of provocative challenge rarely leads to mutual problem solving and more often to dead end arguments.

Avoiding Conflict

Another basic strategy developed by children from families with poor socialization skills is avoidance. Since most confrontations lead to a double-bind where they must choose between total loss of self-esteem or physical abuse, they will opt for avoiding confrontation altogether. Their encounter with either or both parents usually involves situations where the parent carries out an automatic agenda in which the child's feelings are minimized, their arguments discounted, and they are intensely challenged or threatened for displaying any kind of disagreement. They come to see all confrontation and conflict as negative. They observe others keeping secrets or hiding emotions to avoid conflict to keep themselves safe and maintain control. It is only natural to adopt such a strategy themselves. They may even avoid being physically present altogether. Successful conflict where individuals control the level of their anger, argue until compromise is achieved, and then make up, is rarely, if ever, observed. They never have the opportunity to argue successfully themselves and find it only leads to humiliation, defeat, or physical punishment. This is especially true if they are intrusively interrogated all the time, examined, and found lacking or at fault.

These individuals are often quiet, secretive, show little emotion, and are socially withdrawn when it comes to challenging situations. They are fearful of revealing their opinions or their emotions and may constantly be looking over their shoulder. They may be overly self-controlled and overly controlling of others. Since they do not overtly exercise their control, but covertly do it through secrecy, they can conveniently blame others for misconduct.

A No Win Situation

For many of us the world becomes a no win situation as we grow up. We come to assume the worst will happen and that things probably won't work out the way we want them to work out. A few will abandon their faith and allegiance to mainstream social reinforcements altogether and will wall off their feelings in the effort to bury their guilt and self-hatred. Social institutions and their rewards are an impossible no-win situation. Instead, they do what they want and hate everybody for participating in such a cruel and unfair world. They may become the troublemaker in class who sits in the back of the room sabotaging the lesson of the day. They may feel themselves as unredeemable, yet see themselves as having the only realistic perspective while others are hypocrites and liars. They may become conduct disorders and juvenile delinquents. They may or may not get caught, depending on their socio-economic bracket.

What we tend to overlook when we try to rescue or rehabilitate individuals from impoverished socialization contexts and or genetic deficits is the neurological lag that exists. We can temporarily change behavior through environmental manipulations, but without corresponding changes in the neurophysiology little enduring change will occur. It takes time for the brain to restructure itself: to sustain the emission of new, more efficient and more appropriate behaviors. Years of extreme emotional neglect, fear, rage, and hypervigilence alters the physical structure of the brain as well as the neurochemistry. We know this from the work of a host of researchers in the field of neuroimaging such as Drevets (1997), Davidson (2000), Shore (1994). The brain and the mind are disordered. Both the physical structure and the process. There is a clear EEG signature of a disordered mind that shows up in EEG topographic maps (John et al,1988). ADHD, depression, anxiety, impulsivity, etc all have definite profiles which relate to problems in specific networks of the brain and their physical configuration. Individuals with a high level of impulsivity will show excess activity in slow brain wave frequencies in the prefrontal cortex that are two to three standard deviations outside the norm and therefore statistically significant. This means the front of their brain is moving too slow. These individuals are especially responding automatically and with little self-control. This slow wave activity must be reduced through an increased efficiency of the neural networks managing the frontal regions. Only then will they have the inhibitory ability to better control their behavior. Individuals with a history of anxiety and depression demonstrate a loss of neurons and other tissue in hippocampal, caudate, and prefrontal cortical regions (Kaplan, 2002). These areas are critical to the correct functioning of individuals in social situations. These observations are derived from what are termed "MRI volumetric studies" wherein the actual size of brain regions are studied for abnormal variation.

The significance of this loss of structure and function is that stress and trauma alter the brains processing capacity and that time is required for the individual to experience a change in their neural structure (Schwartz, 2002). Long term consistent practice in new ways of thinking, feeling and doing are required. Research suggests that this change can take place as a consequence of cognitive therapy, as Cozolino (2002) has reported in his analysis of the impact of co-narrative exchange at the clinical level, but it is a very slow process. Neurofeedback, a technique involving practice changing brainwave patterns, can accelerate this process greatly because it encourages the individual to directly alter brain function in the areas of greatest deficit (view the impressive list of research yourself at www.isnr.org). By engaging in this type of neurological exercise at the same time as identifying and changing inefficient behaviors, the transformation process can be greatly accelerated.

Summary

Individuals respond to the cues that others emit in ongoing social interactions. In the case of those who are intimates, these responses can become heightened by the threat of abandonment or loss of love and attention. The response to cues is already scripted and is generally set in motion at a very early age in the subcortical emotional circuits, which are faster than cortical circuits, through classical and operant conditioning in the course of socialization (Shore, 1994). The exchange between individuals may be rapid and out of their conscious control as emotional intensity escalates and the ability to utilize the cortex becomes progressively more difficult. The result is often a social exchange that is taking place at the level of a five year old child rather than an adult. The only way to inhibit this process is to rehearse other responses that become routinized and that will redirect behavior and stop the emotional escalation. In a word: training.

The Wisdom of Error and Respect for Ignorance

Making mistakes is crucial to learning. We explore patterns that we have seen or imagine will be successful, we make errors, and we adjust for those errors. We become more accurate in our behavior and consequently more successful in achieving our goals. Since we live in a "social" reality we call this social accuracy.

When we make too many errors we begin to feel more arousal and heightened tension or physical bracing. At first we become more focused and alert as our vigilance system in the brain increases activity and we may become cognitively sharper. If the errors continue we automatically anticipate the negative consequences and may become over aroused. At a certain point in the escalation our arousal will become too high and we will begin to make even more errors. Recovery becomes increasingly difficult as frustration mounts and task failure seems inevitable. At a certain point we walk away, shout, become violent or generate some similar expression of profound frustration.

In many social situations, the longer we can learn to sustain this mounting tension and defer gratification, the more successful we can be. We may build multiple ongoing goals that we work with over long periods of time to achieve complex objectives that provide social rewards others are not trained to garner: such as getting straight A's. This means sustaining mounting levels of tension that we inevitably must habituate to in order to adapt and focus our attention on more immediate tasks. We must relegate the discomfort and tension to background awareness so we can continue to be efficient at our tasks at hand. Every business executive who presides over a meeting knows intimately how challenging this process can be during the tensions arising from complex social interactions where the stakes are high. Many of the executives who have trained with me have also commented on how the whole group calms down when they themselves are less reactive and more centered during meetings.

Managing Error: Mistake Are For Learning

If the cost of error becomes too high, then error becomes increasingly unacceptable. Each error in turn results in higher levels of arousal and increased mistakes. Consequently mistakes become more common. Eventually failure becomes a common pattern. Yet it is dangerously unacceptable. Damned if you do and damned if you don't. This double-bind experience, which R. D. Lang found so common to the experience of individuals with disorders, has been found by Aron Beck (Beck,1995) to be the basis of anxiety and depression. Learned helplessness settles in during extensive double-bind experiences for all mammals in this type of scenario and eventually death if there is no resolution. Individuals give up and wait passively for the inevitable punishment that is sure to follow any effort they make to protect themselves.

Individuals who have experienced this type of situation and recovered are often highly sensitive to any stimulus associated with the experience. They overreact to any mistake or suggestion of error. They live in constant fear of any error because it may return them to that horrific state. They become overly aggressive in all interactions where disagreement hovers. If they are fearful of conflict they will hide their aggression, but the cost is high. If the cost of their past error has been especially high, then PTSD is usually the consequence. Error then results in dissociative episodes where the individual relives the horror of the past or transfers it to the present context, dramatically overreacting to the situation. Grieving is a good example. If proper mourning in the death of a loved one does not take place, then the negative emotions connected with that event may be confounded with future situations where a loss takes place. The individual will not even be aware that they are drawing emotional energy from a past event. This can dramatically interfere with the efficient processing of a current stream of information. The individual cannot effectively solve ongoing challenges that emerge from social interaction. This is not some theoretical formulation derived from Freudian theory, this is empirically derived observations grounded in research.

The Divided Mind

The brain appears to be divided in function across several areas of its structure. Memory involves the entire brain and appears to have a very important primary division. There is a subcortical implicit memory system and a cortical explicit memory system. The implicit memory system is almost entirely below our level of day to day awareness while the explicit system is very much a part of our daily awareness. The outcome of having these two memory systems is very much as Freud suggested. We do appear to have an unconscious processing system that holds memories of which we are unaware.

Emotion seems to issue primarily from the limbic system, an older structure of the brain. The emotional components of memories can apparently become stuck in a state of partial processing if the intensity of feeling is too great. LeDoux's (1996) observation that the cortex speeds up at first to deal with intense activity in the amygdala suggests the possibility that the cortex may become delayed enough in processing that it fails to encode information which is stored in the limbic systems memory. In other words the activity in the cortex distracts us from our feelings. This explains why worry can be so comforting and compelling in a strange sort of way. Frequently we have clients who experience disconnected feelings and body memories that surface in pieces over a period of several sessions. Eventually an image will pull these pieces together into a complete conscious memory that allows the full event to be processed using the cortex. This is consistent with Antonio Demasio's (1999) observation that memory is multimodal and is constructed from many memory centers in different sensory areas of the brain.

Once the memory is integrated into the cortex, the related emotion ceases to be confounded with other events. It also ceases to generate the anxiety that the client experienced whenever an episode threatened to recall it into the conscious memory system. This may be a process that involves amygdalic memories, in the amygdalofugal pathway or implicit memory tracts, becoming integrated through the mammilothalamic/hippocampal system, or explicit memory, into the cortex based memory system. The division in the brain is healed. More to the point, the division in the memory experience is healed. The dynamic tension is released. This process also usually involves intense and vivid dreams, suggesting that a flood of non-conscious processing takes place in a massive integration effort. An integrated emotional component of a memory arises with the sensory memory, binding it, and passes. The emotional component no longer intrudes in an overpowering manner into other novel situations resulting in the elicitation of inappropriate and socially inaccurate behavior. When training individuals with EEG biofeedback we often see a period of intense dreaming coincide with greatly reduced emotional reactivity. Clients comment that things that used to set them off no longer bother them.

6

Digging Out

A house divided against itself cannot stand
Jesus of Nazareth

It has become clear from a great deal of existing research that ongoing social distress, especially in the form of relentless and intense daily hassles, leads to chronic hyperarousal in humans. We tend to habituate to these higher levels of arousal over time and adapt to the consequences. Often these consequences are somatic and surface as physical problems. In fact the American Medical Association, by its own reckoning, has determined that the majority of doctor visits are due to such chronic hyperarousal issues related to social distress. Based at Harvard University, Herbert Benson (2000) has studied this phenomenon for decades and arrived at a simple and elegant solution, he calls it the Relaxation Response. Unfortunately it lies beyond the reach of most Americans for cultural reasons. The relaxation response teaches individuals how to control their arousal level, but most Americans are so hyper-aroused and unregulated that they cannot bring themselves to practice it. They are also so tied to the concept that healing has to take place through the medium of a pill or a machine that they don't believe an alternative can exist. The only individuals who seem able to utilize it are those most desperate or facing death, such as cardiac patients. A series of studies done over the past ten years at Duke University demonstrated that just a modest stress management program can reduce the probability of a repeat heart attack in cardiac patients by over seventy-five percent. That research is being all but ignored by physicians. Our staff presented this and other research as well as a variety of powerful new technologies to address this problem to a local cardiology department at a near-by hospital. They listened politely to the lecture, had fun with the equipment and then quickly forgot the whole thing.

Anxiety Is More Than Worry

It is a common myth that anxiety manifests mostly in the form of worrying. In truth, for most mammals and humans, anxiety tends to surface as physical problems. The first major area of impact for the majority of people is sleep disruption in the form of insomnia. Thirty years of sleep research, from institutes like Cornell University (Maas,1999), indicates that in most cases, insomnia is a consequence of anxiety. It is surprising how many people never connect the two. Many of my students and clients will comment that something troubled them and interfered with their sleep but they do not realize that what they are experiencing is anxiety. They are also not aware of their level of anxiety and how long it has been going on. Sleep is not usually disturbed on a consistent basis unless the anxiety has built to a considerably high level. They have habituated to the ever increasing levels of anxiety and they have not noticed it (remember the frogs?).

The chronic unnoticed anxiety results in a higher average level of adrenaline release into the bloodstream and along with it a substance called cortisol. Chronically high levels of cortisol dramatically impact the brain structures around what is known as the hypothalamic pituitary axis (McEwen, (1987). This group of structures regulate the hormonal system of the body and consequently, sleep, blood sugar levels, thirst, sexual drive, hunger, and immune function. Furthermore, a high level of cortisol destroys neurons in a part of the brain called the hippocampus which in turn hampers short term memory. Volumetric MRI studies show the hippocampus shrinks in size considerably in individuals suffering from long term social distress or trauma (Kaplan, 2002). It is not unusual for individuals with chronically high anxiety levels to suffer serious loss of function with regard to short-term memory. They show up at my office fearful of premature Alzheimers and can't believe it could possibly be stress, until we hook them up to the physiological monitoring equipment. Then they can see directly for themselves how worked up they really are over their everyday life. It is not hard to connect the dots here, but our cultural training prevents us from drawing the obvious conclusion from decades of research done at some of the finest medical institutions in the world. In addition, it is very difficult to confront the fact that you must change your ways.

Biofeedback as a Measure

Biofeedback can be used to empirically measure levels of arousal and often determine which system in the body is most responsive to stress. It appears that individuals differ in their genetic window of vulnerability. We can monitor muscle tension (EMG), hormonal response (SCR), peripheral vascularity (TMP), respiration and heart rate. Different individuals will respond most intensely in different areas of their physiology when distressed. When a client comes in complaining of insomnia, we frequently hook them up to the biofeedback equipment for a demonstration. I ask them to completely relax and then show them their level of arousal based on physiological measures. Many are surprised to find that their autonomic systems are running at a high level even though they feel relaxed. What is even more surprising to most is that, try as they may, they cannot reduce those levels of arousal. They can't really relax.

What we can do with biofeedback is feedback physiological information about their high levels of arousal to them in visual and/or audio information and have them practice using breathing and visualization to gain control of their autonomic systems. As they become aware of their bodies they are often surprised at how much discomfort they are in. They may complain they are feeling worse, rather than better. This can often become a roadblock to their progress, but if they are willing to relax enough to become aware of their actual level of physical discomfort, then they can often learn to relax enough to release the tension generating the discomfort.

Occasionally, a client will relax into a mild panic attack. This may be presaged by growing levels of anxiety or a sense of falling. We call this abreaction. People tend to brace themselves physiologically, tense up, to hold back emotion and when they relax that bracing, the emotion is released. Often this may be in the form of shaking or crying. This is very interesting because Joseph LeDoux (2002) has observed that freezing and shaking are natural ways of coping with danger for most mammals. The amygdala, the fearful watchdog of the brain, has direct pathways to networks that control these functions. In fact many animals will die if they do not shake enough after encountering danger. Humans seem to have the added abilities of crying and laughing. These are also potential modes of abreaction that we observe occurring in our practice.

Neurofeedback

Neurofeedback, also known as EEG biofeedback, is a form of biofeedback wherein brainwaves are used as a basis of biofeedback instead of traditional measures of electrical monitoring of the biosystem, such as EMG or muscle tension, that monitor the body. In this case we are measuring electrical activity in the primary regulator of the Central Nervous System, the brain itself. We can feed this information back to the individual, once it has been properly analyzed by a computer, in a manner that allows him or her to become aware of their brainwaves and control those brainwaves. The consequence of altering the level of electrical activity in a particular area of the brain, is that it alters blood flow as well as neurotransmitter activity. By training individuals in this skill, we are able to permanently alter their neural activity. The result is that we can effect changes similar to those generated by psychoactive pharmaceuticals but the individual does not become dependent on the training to sustain those changes.

This fantastic sounding claim is not so difficult to accept once we realize that most mental disorder is a deviation from mental order that has a fairly specific profile or signature. The brain is structured to work best within the parameters of this profile and consequently prefers this pattern of functioning. As James Austin (1998) has commented, the brain is in fact delicately balanced between coma and seizure. It can apparently grow in a manner so that it achieves an adaptive homeostasis, or balance, in order to deal with stress, but it is not its preferred mode of functioning. Ron Ruden of Mt. Siani hospital has termed this neo-homeostasis (1997;1999).

This phenomena can be likened to a spinning top that has been bumped and begins to wobble but still maintains its balance. This wobbling is related to slowing from its ideal speed of functioning. The brain also appears to speed up or slow down in response to stress. If it becomes stuck operating at either extreme state, then aberrations in function begin to occur. Noise becomes introduced into the system. It is similar to having your computer play tricks on you when there is not enough memory or when it is not operating at the right clock speed. The preferred operating frequency is what is called an attractor state in nonlinear dynamic theory. It provides what might be called an optimal coupling zone for brain networks to function in (Nunez, 1995).

We have documented that individuals with anxiety tend to have a nervous system that runs too fast and that individuals with depression tend to have a nervous system that runs too slow. Those coming in with anxiety disorders have brains that are overactive and give off an abundance of high frequency brainwaves. Those with depression or ADHD have brains that are underactive and emit too much low frequency brainwave activity (John et al, 1988). Neuroimaging supports these observations. The existing PET scans of these disorders show clear evidence of deviation from normal levels of activity as do the qEEGs, or topographic brainmaps of electrical activity, we have done of our clients. These deviant levels of brain function result in equally deviant levels of activity related to thinking, feeling, and doing. When we utilize drugs to adjust for these difficulties, we are adjusting neuromodulator systems that alter the responsivity of various areas of the brain to neural resonance via the thalamus and consequently the efficiency with which certain circuits operate.

David McCormick (1999), one of the leading brain researchers in the country, not long ago published an article in which he argued that the disorders we deal with in the brain may be largely a consequence of thalamocortical dysrhythmia, or in other words a lack of self-regulatory control in the brain. Others in the field have begun to favor this argument because of all the supporting research. Another more recent study by Stuart Hughes and Vincenzo Crunelli (2005) also strongly supports this theory. In EEG biofeedback, or neurofeedback, it has long been accepted that this is one of the primary areas of the brain we are impacting. Barry Sterman (1972;1986), a neuropsychologist at UCLA, has thirty years of high quality research that clearly demonstrates this; mostly ignored by the medical community. With EEG biofeedback a researcher can even teach a brain to stop having seizures. Now that's self-regulation!

Inducing The Relaxation Response

Herbert Benson (2000), a researcher at Harvard University, studied individuals in various forms of meditative states, starting in the 1970s, and found they had several things in common. Of particular interest to him was that they all entered a profound state of relaxation he described as a hypometabolic state. The term was used because their metabolism was even slower than those of individuals in non-REM sleep. He published his findings and officially designated this state the Relaxation Response. Benson found he could teach the average person to enter this state using a simple generic nonreligious technique. When he and Wallace (1970) taught individuals with cardiac problems this technique it greatly improved their health and recovery. Benson went on to apply this technique to other problems and found it could help people greatly with variety of problems. It was especially effective with anxiety and related psychophysiological disorders involving chronic hyperarousal. For those of us involved in biofeedback, this was very good news because we had the equipment to monitor individuals and teach them to enter that state. All we needed to know was the basic indicators that the individual had entered this state and we could guide them using modern technology.

You might ask, "Why bother with the technology if Benson had developed a good technique?" This brings us back to the problem of cultural bias and training. Most westerners don't learn this technique easily and often do it improperly or give up too soon. It is harder to do than it first appears to be. The technology allows us to guide someone very specifically in entering that state and allows us to monitor them to make sure they sustain that state. It also provides a concrete external measure of performance that westerners seem to need. Finally it provides a technological solution in a technologically oriented culture. It increases the efficiency and effectiveness of this technique.

Over the last forty years many researchers have looked at the physiological and EEG correlates of meditation (Austin, 1998). What they have found is that meditators generally display similar specific measures of physiological change. Their breathing slows from a norm of around 16 breaths per minute to six. Their heart rate slows. Their breathing becomes softer and more abdominal. The CO_2/O_2 ratios change as their exhalation period increases with respect to their inhalation period. Their breathing rate alters their physiology dramatically. Galvanic skin response drops, which means their autonomic nervous system becomes very quiet and muscle tension, or EMG, drops to a very low level. EEG recordings indicate they generate high levels of alpha brainwaves especially in the central and frontal regions of the brain. The specific frequency varies within the alpha range depending on years of practice, but it is the dominant spectral component. The research so far mentioned very clearly demonstrates that meditation is a very different state than relaxation and drowsiness and Dunn et al (1999) have specifically addressed this issue and demonstrated this difference in their research.

Brainwave Frequencies

Before discussing anything more we need to brush up on EEG or brainwave theory. Alpha brainwaves are smooth sinusoidal brainwaves that come and go in bursts known as spindles. These brainwaves dominate the brain when we are relaxed and engaged in some familiar repetitive activity. This could be knitting or it could be shooting hoops. It could even be playing a well rehearsed musical piece. Alpha waves disappear when we start to work on a serious problem in our minds or when we worry about how to arrive at a solution to a dilemma. At the same time a faster frequency of brainwaves, called beta, appear and they indicate a lot of neural processing is going on in the brain. It must be pointed out, however, that some people are so repetitive and accomplished at worry that they produce a form of fast alpha instead of beta. Another common form of brainwave is theta and it appears a lot when we are remembering, daydreaming, or drifting off into sleep. Finally the fourth most common category of brainwaves, and the slowest, is delta and it usually dominates the brain when we fall asleep. From all of this we have observed that people are shifting in and out of different brainwave categories minute by minute as they go through their day. However, people in unusual brain states or with disorders may tend to dwell in one category more than another. They lack neural flexibility or plasticity. Some individuals cultivate particular categories of brainwaves as a consequence of certain exercises and one such group is individuals who pray or meditate a great deal.

Research on The Relaxation Response Continued

Two investigators in Japan, Kasamatsu and Hirai (1969) investigated Zen masters and their students during their meditational exercises. What they found is that there was a dramatic increase in brainwaves in the alpha band initially and that over time possibly even some frontal theta began to dominate in advanced meditators. Their EEG amplitudes were very high, 60-100uvs, and increased in relation to number of years in practice. The average person has alpha in the 12-18uv range and it rarely goes above 50uv (Neidermeyer, 1999) so you can see this is quite dramatic. Their ability to produce slowed alpha was also correlated to number of years of meditational training. Maxwell Cade (1987) found the same pattern with meditators in England and added that he noted that, unlike the alpha slowing that occurs in individuals with depression which is asymmetrical, the meditative alpha was balanced in both hemispheres. Kasamatsu and Hirai also found that the high-amplitude, high-coherence alpha became dominant around the central and frontal region of the brain. This is another way of saying the whole brain was doing alpha at the same time together. A large number of neurons were synchronized together.

Research on Transcendental Meditation (TM) showed increases in meditators in alpha as well, plus increased coherence, which again means the whole brain was greatly involved in this process (Banquet, 1973). Recent Research using qEEG brainmapping techniques revealed a very similar pattern of increasing alpha amplitudes in mindfulness with greatest amplitudes in the central region (Davidson et al 2003). However, these were not advanced Zen Masters, which may explain the lack of report regarding frontal alpha. The greatest amplitudes in beginners in concentrative meditation was in the parietal region (Dunn et al, 1999). Other recent work by Andrew Newberg using PET scan technology indicated that this area was also highly active in the individuals he monitored during peak meditational experiences as well. He hypothesized that this area was especially inactive because it was the section of the brain that worked to constantly maintain the physical boundaries of the body, of self from not-self. Consequently, the sense of self was diminishing in an experience of merging with the other. From an EEG perspective this makes sense as well. There is a 70% correlation between EEG activity and blood flow in the theta and alpha ranges of activity. The slower the activity, the lower the perfusion of the area with blood. High amplitude alpha in that area would indicate lower metabolic activity because circuits are in an idling mode.

So, we have extensive research showing us the basic indicators of the meditative state as well as a generic technique developed for inducing it. Individuals can be trained in the technique and monitored. Any area where they are not performing properly in can be noted, assessed and compensated for through biofeedback training. Interestingly, very little of this information is being used in our culture, except at our clinic (and as far as I know Les Femi at Princeton (1998)) and we are using it very successfully with mental disorder as well as with peak performance.

The First Effects of Training

When we begin training individuals to do the relaxation response at our clinic we often monitor multiple areas of physiological activity. We initially train them in breathing techniques. We also monitor peripheral vascularity and muscle tension to see if we need to assist them in these areas. In biofeedback you quickly learn that people can store their tension in different physiological areas. An individual may have relaxed muscles, but their SCR readings may be swinging wildly up and down. Our goal is to make them aware of this dimension of response to stress and identify what in their life provokes it. This is part of their self-awareness training as well as transformational training to prepare them for transcendent experience. Without taking action to resolve their attachments in specific areas of their lives, they will not be able to successfully engage the meditational process.

Next we begin giving them feedback whenever they are producing alpha. Using specific techniques and protocols we progressively train them into deeper levels of the Relaxation Response. For some periods we provide feedback and for some periods we merely monitor their activity. As individuals begin to gain control of their EEG and clear their minds, they begin to experience specific effects. For healthy individuals these effects occur very quickly. For those with long standing disorders the effects are more pronounced and appear slowly over time. Usually they begin to sleep better and have vivid dreams. Often they find themselves drawing boundaries more clearly in their everyday lives. They become less automatic and more innovative in their response to difficult situations. They experience an increase in energy and more positive moods. They often report that their "buttons" are not as easily pushed. These are the signs of integration and growing self-awareness. In time this self-awareness will mature into deep insights regarding their own behavior and a growing liberation from deep seated beliefs that walled them off from transcendent experience.

Abreactions

Some clients have braced and struggled for so long to keep their feelings at bay, that relaxing their mind feels very threatening. Often they cannot stand more than a few seconds on the entrainment devices we sometimes use to initially introduce individuals to an alpha state. These devices gently pulse light into the optic nerve and generate a resonance in the cortex at the alpha frequency. It provides a preliminary subjective experience of what alpha feels like so that individuals can identify the state easier when they are doing neurofeedback. Sometimes a preliminary introduction to this state is too much for some individuals and they begin to experience nausea, sweating, or panic. This is usually an indicator that they have some buried trauma that has not been addressed. This must be dealt with before they can achieve access state. This is not based on any clinical consideration, it is just a fundamental fact in our experience that these individuals simply cannot engage the Relaxation Response without remedial work. When they do neurofeedback, they begin to panic as well. As they relax into their body and their feelings, they become overwhelmed with suppressed thoughts and feelings; and perhaps memories as well. Some have been classically conditioned through traumatic socialization environments to remain hypervigilant in order to remain safe. To relax means to subject themselves to danger. Consequently, the neurofeedback process can become threatening in itself and they may display reluctance to come in for training. This demonstrates why preparation for meditative experience is as important as the meditational training itself.

Long term effects

As I mentioned earlier, over time, as training proceeds, clients begin to sleep better. This is often the first indication that they are healing. For most people, the first signs of anxiety are difficulty sleeping and an attendant loss of awareness of dreaming. Research indicates that 95% of insomnia is due to anxiety. However, many people tend to somaticize their anxiety, that is to say they experience their anxiety physically as a symptom rather than cognitively and emotionally. For these individuals the insomnia may not occur. In fact they may not feel anxious at all and deny it is an issue. They will even look good when tested with psychometrics. For those with sleep problems, when sleep improves, their REM and delta sleep patterns normalize. This results in more dreaming and more awareness of dreaming. As a consequence of this, their attention and cognitive processing as well as their memory improves. Problem solving ability improves and they are able to gain insight into their problems and possible solutions as well.

As they begin to become more aware of their problems, confront them and solve them, clients begin to alter the pattern of their social lives. They tend to engage in approach behaviors instead of avoidant behaviors. They participate in social interaction more and do a better job at negotiating their identity and accessing the social resources they require. Their behavior becomes less deviant in many spheres where it had become problematic for them. Isolation, deviant behavior, and personal suffering decrease. They begin to feel more powerful.

As confidence increases they begin to draw personal boundaries better. They take back their personal power by refusing to let others dictate and control their feelings and actions as much as in the past. They are more aware of their feelings and they are more likely to honor them and act accordingly. Many learn to say "no" where previously they always said "yes" even though they could not afford to do so in terms of their own resources.

The Research on Sleep and Cultural Consequences

Most Americans are sleep deprived. On average, researchers tell us, most people got nine or ten hours of sleep until the invention of the light bulb (Maas,1999). This allowed production to continue on longer into the night and allowed people to stay up later with more satisfactory lighting. This extended opportunity to produce, but in a society that was built upon the Protestant Work Ethic, it opened the door to new forms of subtle abuse and exploitation. Where previously individuals were limited in their production efforts by daylight, now they were unlimited. Employers could force their employees to work three shifts around the clock and cite traditional sources of moral justification for hard work, when in fact it was just greed that motivated them. With the added weight of policy and legislation to coerce labor into this mode of operation, the whole society rationalized the concept that there was "no reasonable limit" to how much work one should do as well as the idea that the person who worked harder and longer was morally superior to others. Of course management did not realize it was writing its own obituary as well. Today executives suffer under the same phony moral imperative. Presently it is destroying their health and sapping their resources just as much as anybody else in the workforce.

In spite of the fact that Americans outwork any other culture in the world, we still exhort each other to work harder and achieve more. People have come to believe that six hours or less of sleep is enough. For many it is something to brag about and a means of demonstrating how tough they are compared to others. This is true in the short run, but in the long run it takes a serious toll on their health. Memory problems, poor concentration, high blood pressure, anxiety, depression, overeating, dependence on alcohol, and a long list of other ailments related to sleep deprivation are considered to have magically appeared in their lives. They will complain of bad luck or poor genes. And of course the solution is more drugs to deal with the symptoms. This elaborate system of denial allows the abuse to continue. A negative cultural institution that supports many industries. Our productivity may be high, but so is the long term health cost of maintaining it.

Our relationship to work and sleep is an example of many other unexamined dimensions of our culture and the lifestyle that emerges from it. It is very instructive and demonstrates how we institute denial and justify destructive behaviors such as greed. It shows how automatically we adopt the cultural standards. When clients see how they have allowed themselves to be enrolled in this type of thinking that leads to self-abuse, they begin to alter their sleep habits and experience first hand the amazing changes that take place. In conjunction with the neurofeedback they begin to regain their energy and positive mood as well as their memory and concentration. They also begin to experience first hand how automatic they were in their beliefs and behaviors and how like a trance their daily life pattern had become. This is the cultural trance. This is how as a group our individual automatic behaviors merge and become external and coercive social forces that will define future socialization processes. Self-awareness is the only solution. Neurofeedback is an excellent method for training self-awareness.

Alpha Training

In training the Relaxation Response with Neurofeedback, one of the key frequencies reinforced, as you can image, is alpha. This is clearly because it is the frequency most commonly seen in advance meditators. Alpha coherence and synchrony, as we mentioned, is also an important dimension of training. I like to compare the brain producing alpha to an engine idling. Most healthy adults produce alpha between 9.5 and 10.5 hertz. Depressed people and individuals suffering from toxic encephalopathies (poisoning such as from mercury) generate a slower alpha (John et al, 1988). In fact slowing is an abnormal condition of the brain. So is too much fast wave activity. This shows up in people with anxiety. The brain idles at 10hz and moves in and out of this frequency range as it performs various functions. When the brain shuts off for sleep it tends to go into delta.

Barry Sterman (1995), working with the military at Edwards Air Force Base, found individual pilots who performed best did so because they were able to return to alpha more frequently while on task than pilots who were not. This indicates that the brain in its optimal state of functioning needs frequent relaxation and recovery time. Of course we note this with neurons as well. They have a recovery time after they fire an action potential. Muscles as well require periodic recovery time or they build up too much lactic acid. It is part of the biologic imperative to function rhythmically.

Joe Kamiya (1969) found as far back as the late sixties and early seventies that individuals can reduce both state and trait anxiety by increasing alpha amplitudes for extended periods of time through brainwave training. In conjunction with biofeedback this process is even more effective. This correlation between alpha production and CNS arousal reductions suggested that alpha could be a measure of CNS relaxation or the Relaxation Response as described by Herbert Benson.

Many early pioneering trainers were not especially careful how and where they trained alpha in the brain and this of course lead to very mixed results. It also lead to an oversimplification of the entire process. As with many discoveries, there was great initial excitement and wild speculation as people hoped for instant enlightenment through alpha training. This response is very typical of our culture, which likes instant everything. Of course when it turned out not to be neither instant nor dramatic, bitterness and disillusionment followed. People assumed it did not work at all and went in search of the next immediate gratification.

There were many in the field, however, who saw the value of this approach and continued to work patiently and scientifically with it. Les Femi (1978;1980), at Princeton, in particular, spent decades integrating research on attention and alpha synchrony documenting its robust and useful effects on human consciousness. Adam Crane continued to successfully promote and employ alpha training in the healing and recovery of individuals with a wide variety of disorders as well as peak performance training. Max Cade in England did extensive research and development on meditative states and alpha training. The alpha-theta training techniques developed in the late 80s by Elmer Green (1977) were used very successfully with alcoholics and war veterans with PTSD at the Menninger Clinic in Topeka Kansas by Eugene Peniston (1989). These studies were published and replicated. Bill Scott followed with a series of experiments in this area, the largest and most successful of which was just conducted through the UCLA department of psychiatry. John Gruzelier (2003) in England has recently conducted a series of controlled group designs. In fact, within the field of Neurofeedback, there are probably more controlled group designs conducted and published on alpha training than any other approach. Yet it continues to be denigrated by many in the field who are poorly informed. We will explore how this came to be next.

One Step Backwards: Poor Science

Early in the research on alpha some skeptics did some experiments which suggested that Joe Kamiya's research on how Neurofeedback can reduce anxiety was poorly done and that individuals in dark rooms could produce the same amount of maximum alpha as brainwave training. Unfortunately they did not fully understand the dynamics of alpha or EEG biofeedback at the time and ended up creating a design far worse than anything Kamiya could have devised. Their findings, although highly biased and spurious, were published in a high profile scientific journal and nevertheless remained the final word on the subject. Many in the field of neurofeedback who also poorly understood research methods and were easily swayed by authority rather than science began to ignore alpha training. They perceived it as an early mistake in the field's development. This kind of political bias and arrogance, sadly, is not unusual in science and often leads to set backs in many fields (Kuhn, 1970).

Meanwhile, as we mentioned, pioneers in this area who realized the flaws in this research continued to explore the potential of this approach. Bill Scott, Les Femi and Adam Crane as well as myself have continued to take the lead in this area. Joe Kamiya (1969) has continued to encourage others to read his research more carefully and follow up with their own. Anna Wise (1997) also developed the utilization of alpha in her work as did Maxwell Cade. Unfortunately the first book published about the field of Neurofeedback Symphony in the Brain (Robbins, 2000), missed this entire dynamic as it focused on only one group of clinicians on the west coast. I have discussed this issue with the science writer Jim Robbins who wrote Symphony of the Brain and he acknowledged this omission in retrospect. He has recently gravitated toward working with Les Femi and is beginning to explore this overlooked area of the field as well.

My own theories regarding this issue over training alpha were first presented in the book Adam Crane (Soutar & Crane, 2000) and I did together called "Mindfitness Training." Adam has based his whole mental health and optimal performance seminars on this perspective, which he has been pursuing on his own. Adam is somewhat of a visionary and intuitive and has been fitting the pieces of the puzzle together in his own creative way. When we met, we were shocked at how similar our own lines of thinking were with regard to this phenomena. Since that time the research has moved consistently in a direction which supports this perspective. Our clinical work and qEEGs clearly reflect this pattern as well.

The brain maps we have done on almost every existing mental disorder show a deficit of 10hz alpha and and/or an elevation in low or high frequency activity. This means there was a lack of the usual dominant alpha rhythm and too much EEG activity in the slow or fast frequencies. This was not at first apparent from the existing brain map or neurometric databases because they did not break frequency bands into fine enough single hertz analysis. Generally they would show the activity in broad bands such as alpha bands from 8 to 12 hz only. Many people with depression would show high central alpha and many in qEEG simply assumed this disqualified the theory that alpha was generally a good frequency to train. They would say "One size does not fit all." Of course this ignores the fact that we all have a three pound brain that does fit in our skulls and that they all work best around 10hz.

So, along comes Bob Thatcher and Bill Hudspeth with this much more detailed analytical database with 1hz bins, that is to say the maps show how much normative activity there is in each frequency band. It turns out that the high amplitude alpha in most pathologies is slowed alpha or fast alpha. I had been predicting this pattern for some time and many associates in the field who attended my workshops were skeptical initially. It turns out, however, that excess 10hz alpha is very rare, and usually occurs in areas where it is not supposed to occur and not with the same coherence and phase patterns as meditators. Attending this shift in alpha to a higher or lower frequency domain is excess brainwave activity in other domains. Individuals may have a brain running too fast with excess beta or a brain running too slow with excess theta and/or delta. Sometimes they may have both at the same time. In each case different disorders emerge. The symptoms of these disorders can be diminished by retraining the brainwave patterns. This suggests that meditation is a special state reflecting an extra-ordinary balance of activity in the brain.

The research on meditation reveals a very specific pattern and direction the brain takes during each phase of the process. Research by Richard Davidson (Goleman, 2003; Davidson, 2004) as well as John DeLuca (2003) and myself (Mindcycles, 2004) all demonstrate that the various meditation techniques have unique neurological signatures that can be reproduced by different individuals trained in that specific technique. The Relaxation Response is a special type of meditation known as concentrative meditation. It is the technique used by Buddha, taught in Zen, used in Theravadan tradition, discussed by the famous Yogi Patanjali, and taught by the Maharishi. It is the technique used to attain access to the different levels of consciousness unique to these traditions called Janas. By using brainwave training we can help westerners achieve this state more quickly and effectively so that they can engage these unique states of consciousness for their own personal growth and development.

Brainwave Patterns of The Relaxation Response

The alpha pattern that emerges during meditation is different from other states we usually pass through. Initially the frequency increases to 11hz and then begins to diminish while amplitude increases. Very high levels of amplitude occur in seasoned practitioners as they learn to get more and more of their cortex resonating at the alpha frequency. This results in high levels of what is known as coherence as well. Coherence reflects resonant activity in the brain. This high level of alpha resonance occurs especially in the front of the brain. As practitioners become more proficient in their practice, the research shows that their frequency goes lower (Cade 1987; Kasamatsu and Hirai,1969). If they lose the thread of concentration, as even seasoned practitioners do on occasion, their frontal alpha synchrony drops out and they briefly slip into the delta frequency and unconsciousness (Banquet, 1973: Westcott, 1973). This high amplitude resonance, or synchrony, in the cortex allows them to remain alert and quiet as they drop closer to sleep but never quite enter it. They begin to experience more fundamental states of consciousness that are lost in the chatter of everyday brain activity. As they enter into these more basic layers of consciousness, as Antonio Demasio calls them, they begin to have a more unitive experience of perception. This fundamental experience of unity causes a radical shift in perspective and attitude. The processing illusion of self and other becomes transparent and the tacit experience of the unitive field of existence becomes a reality rather than a metaphor. It is a moment of transcendent experience. The impact on psychological integration is profound. The experience of fulfillment is complete.

From a scientific perspective I would propose that this is the ultimate maturation experience of the human organism. It is the flowering moment of the psyche, and it is a rare bloom. To achieve this level of experience, I would argue, requires both a high level of Self-Integration as well as a high level of Social Maturation. It cannot be done without a great deal of self-regulation skill as well as the support of a well developed social order. Most individuals who engage this type of experience, contrary to their occasional claims, live under the protection and support of a well developed social order. They practice and study continually, year after year, to attain the levels of insight necessary to arrive at these types of experience. It is clearly not the type of achievement that one attains with a few meditation lessons, a short workshop, or even a few years of dabbling.

7

Waking Up

Nothing changes until it becomes what it is.
Fritz Perls

The spiritual side of training is waking up from our automatic nature. Gurdjeff made much of this in his esoteric school of training. It is a key theme of Hindu tradition as well. The word *Buddha* means "awakened one." Tibetan Dream Yoga focuses on waking up in your dreams. The Hindu religion insists that we are living in illusion. Wolinsky (1991) has explored the psychological side of these traditions in his book Trances People Live. In his clinical experience, most people with serious problems spend a great deal of their time in trances. A key feature of most mental disorders is dissociation. Those of us who are not heavily involved in this process are nevertheless constantly overwhelmed by it. Our culture promotes dissociative states. A great deal of recreational activity involves dissociative states. Although they can be pleasurable, they can also be very negative and destructive. Advertising stimulates them in us. Listening to conversations, stories, or watching television induces them. They often motivate us to high levels of attainment. When they are negative and overwhelm us they can plunge us into our own personal purgatory. The relaxation response and concentrative meditation teach us to become aware of dissociative states and to let go of them. It gives us power over our *Samayama* as Patanjai (Isherwood, 1969) calls it. We gain power over illusions. We are able to live in the here and now and it becomes more powerful the more we practice. And it does require constant practice to attain this level of self-regulation.

Dissociation

Clients, as I have said, constantly dissociate into fearful thoughts and images. The intensity of this dissociation can vary as can the length of time. We find that most of our clients are unaware that they are doing this. In fact research indicates that most people pay little attention to this process although they do it all the time. People tend to daydream about success and romance and sex quite a bit. They may shift in and out of these segments constantly all day long. They may move into internal dramas regarding a problem they are trying to solve, trying out different dialogues and scenarios. Many are constantly reviewing their "to do" list. These shifts generate emotions and physical feelings that in turn draw up other images. This interaction with worlds of feelings and images is constant. In fact it is impossible for most people to stop this internal dialogue even for a moment.

The nature of this internal dialogue can have a profound impact on us. Aaron Beck's (1979) research shows that individuals with depression and anxiety tend to engage in constant negative dialogues centered around forgotten conclusions or core beliefs regarding safety and self-worth. We tend to habituate to these patterns of internal response in the same way that we routinize and habituate to other patterns of behavior. Consequently we may develop long-standing habits of dissociation, which interfere with our attention to incoming stimuli and the interpretation of those stimuli. One fearful cue may send us off into a dissociative episode that brings on powerful feelings of anxiety or shame and as we return our attention to refocus on the stimulus we may forget what we have just done but still have the feelings. Those feelings can then become associated with the new stimulus as well. Through the process of classical conditioning we may develop whole aggregates of relations to social objects that are shaded with negative emotions. As these aggregates build through this snowballing negative conditioning we may move from anxiety into functional depression and then deep vegetative depression. This process is slow and varies a great deal from person to person depending on their resilience due to genetic and socialization factors. It may take a couple of years or it may take decades. The constant appearance of negative states will generate cortisol at a level that will eventually erode the hippocampus and other key regulatory areas of the brain such as the hypothalamic-pituitary axis (McEwen, 1987). The gradual nature of this process tends to make it invisible to most individuals in our culture and so they tend to ignore or discount the impact of their poor cognitive-emotional habits over time.

Breaking the Trance

The process of alpha training and meditation encourages individuals to constantly renew their vigilance and break free of this conditioning by refocusing on the same stimulus over and over again. By doing this individuals become sensitized to their dissociative process. This can become uncomfortable because they become more aware of the negative thoughts and feelings that had been automatic and routine. Beck refers to these as automatic thoughts. The advantage with using neurofeedback initially in this training is that it provides a more powerful stimulus to focus on for a modern individual than traditional icons. It also gently forces the individual to produce alpha more consistently through the process of operant conditioning. It is precisely this alpha renewal process that provides the passive attentional state necessary to become acutely aware of internal stimuli. Consequently it accelerates the learning curve greatly.

During neurofeedback training the individual focuses on the same visual and auditory tone or pattern for thirty minutes. Each time the brain goes into the right type of alpha state, the pattern indicates to the individual that they are focusing properly. At the same time this is happening the individual is getting distracted by the usual daily automatic internal chatter or internal stimulus. By becoming aware of the internal stimulus and then breaking away from the neural patterns it generates to reengage in the alpha stimulus, the individual becomes progressively skilled at developing a strong alpha attractor state that involves their active and automatic attention. At first this requires great effort because the established habits of dissociation are very strong and continually pull us into "discursive thought" in which we lose our self-awareness. With practice, the effort becomes more natural and easy as well as habitual. Since alpha is the natural resting frequency of the brain, it is eventually easy for the brain to engage in this process. Individuals spend progressively more time in this alpha state and less time dissociating. As the networks of dissociative episodes weaken from this fragmenting process, the individual becomes more focused in the present and the stimulus it provides. This is the process of detachment and deconditioning that the traditional eastern literature refers to. Individuals give up their attachment to both thoughts of pleasure and fear. This tends to inhibit amygdalic networks that cue the brain about environmental danger. Our clients find themselves less reactive to what was previously aversive social stimuli, especially internally.

A side effect of this process is that neurochemical changes take place as well as physiological effects associated with the relaxation response. Neuromodulator circuits become less activated and stressed. Cortisol levels drop. Some investigators report higher serotonin levels (Austin, 1998; Rudin, 1999). Circuits that have adapted to prolonged distress by adjusting their synaptic receptor levels tend to recover in terms of receptor population levels and sources of transmitter available. There is evidence that the absence of distress can result in neurogenesis, the spontaneous regeneration of neurons, in the hippocampal (memory) regions as well as the prefrontal cortex (executive brain) (Kaplan, 2002). Past evidence that hippocampal changes related to neurogenesis, or the regeneration of cells, can result in increased immune system function has been recently supported by research done by Davidson et al (2003) on immune function and mindfulness. The work of Jeffrey Schwartz (2002) at UCLA also demonstrates changes in caudate structure as a consequence of mindfulness training has been verified through volumetric MRI studies. All these studies support the hypothesis that using neurofeedback to recondition the automatic brain can highly influence key areas of the brain involved with our response to stress and the damaging consequences prolonged exposure to stress can have on us.

The work by Schwartz has been very instructive as it investigates what has been identified as the "worry circuit." Although the malfunctioning of this important region of the brain that is an interface between emotion and cognition is marked in individuals with OCD, it is the same region that is overactive in lesser disorders relating to everyday dissociative processes. The fact that the structure and function of this region and supporting networks can be altered through an attentional training process such as mindfulness suggests the power of this type of training. Alpha training can be very specifically designed to enhance the efficacy of this type of training and has been used very effectively with OCD at our clinic.

Patanjali & Advanced Meditative States

The Yoga Sutras provide an interesting commentary that runs parallel to what we are finding through the neurofeedback process. The NASA physicist Elmer Green became aware of this parallel and discussed it extensively in his book Beyond Biofeedback (1977). The reputed author of the Yoga Sutras, Patanjali, describes different states/stages of consciousness or functioning associated with the task of achieving Nirvkalpa: "awakening from delusion". The process involves developing one pointed attention until a state called "dharana" is achieved. This is the ability to maintain a certain level of attention associated with a profound quieting of the mind. He then describes a series of successive stages through which the individual passes in which the thought stream or internal dialogue is absorbed progressively by one repeating thought wave. Eventually only this one thought wave remains. Finally, as this final thought wave is released a profound new state of awareness is achieved. Adam Crane (2000) has appropriately termed this state of neural silence "Profound Attention." These advanced states are also described in similar fashion in other ancient texts in Buddhism and a variety of other religions. Meister Eckhart, a more recent Christian mystic as well as the more ancient Desert Fathers, describes in detail many of these states (Goleman,

1988). From the various descriptions, it appears they consistently involve this state of neural silence as a prerequisite.

These more refined states of higher awareness are ultimately the goal for these traditions and are beginning to be studied with our modern neuroimaging technology by Davidson at Keck Labs at University of Wisconsin (Goleman, 2003). John DeLuca (2003) has also done some pioneering work to demonstrate that each of the meditative stages has a unique EEG signature across subjects. Our present knowledge and technology, however, limits our ability to engage in this research. Eventually we will find out more but there may be a limitation to our ability to develop indicators of these finer and more subtle states of meditation. Once access state is achieved, these states become far more accessible to the western mind. The key difficulty at this point in our development as a culture is achieving access state. We have the technology now available to move in this direction and this present text is dedicated to defining that now measurable and accessible path as well as distinguishing the problems in achieving it and describing how to overcome these problems using our present technology.

Daniel Goleman (1988) in describing the progression of states of higher awareness in Buddhist technologies of awakening has developed a very appropriate western term for achieving the first level of meditative skill. The initial state in this paradigm, which is parallel to Patanjali's dharana, is called "access state." There are then a series of progressive states of developing neural silence that he describes in modest detail. As this type of training progresses there is a breakthrough in the last stages into a profound experience of awakening.

Insight

It is easy to get overly focused on altered states of consciousness and bliss in the process of pursuing awakening. We tend to focus on the technique of achieving special beneficial states and overlook other factors that indirectly contribute to the achievement of those states. There is presently a trend among clinicians, coaches and business trainers to train people to get into the zone or a state of "Flow"(Csikszentmihalyi, 1991). In these optimal states individuals will be able to achieve new levels of performance. In a society based on meritocracy, performance is the key to achieving high states of status that allow us to access all the best social resources. The idea is to adapt advanced states of consciousness to this process and harness them to move us into higher states of achievement in our careers. The idea appears to have grown out of the martial arts to some degree wherein these meditative states are combined with physical training to achieve super human abilities. The idea also flows from Zen to some degree in that we have been informed as a culture that Zen states can be applied to everything (Watts, 1961). There is also the intriguing notion that if we can just get the formula right, perfection in all things will flow. Stories of sudden enlightenment and salvation also feed into this mindset. Much of this overlooks the fact that stories of sudden enlightenment usually involve characters that have sacrificed everything, including years of their lives, in a search that builds slowly to this dramatic moment. In a culture that dwells on instant gratification it is easy for people to overlook the long process that leads to the rather rapid conclusion. Cultural bias tends to distort our perception in this regard.

Many of us tend to conclude that awakening is merely a blissful state of consciousness like something an individual would experience on LSD. In fact two Harvard psychologists came to that very conclusion in the 60's and arrived at a dead end in their progress. One of them, Richard Alpert (1976), went on to investigate the mechanics of meditational practice in India as a means to achieving similar states of consciousness as he had experienced with psychedelics. In looking back, Alpert, or Ram Dass as he has come to be called, talks about his constant focus on avoiding being "brought down" from his meditational states by stimuli that were aversive to his efforts i.e. certain persons, places, or things. He observed with some humor that this perspective cut him off from experiencing everyday life and was a form of attachment to the pleasure that these higher states could generate. This attachment was just another roadblock to his development on the road to transcendence. Eventually he concluded that there was more to the process than achieving and holding on to the blissful states of consciousness he was experiencing.

Other westerners investigating this process discovered the same thing. Spending years studying with accomplished masters in the field, they found that ultimately insight was the most important component. Most schools of meditation also insist that students study texts and take up practices exercising their insight. Insight meditation is a process whereby one takes the attentional discipline developed in meditation exercises and applies it to consideration of a topic such as developing compassion for others (Kamalashila, 1995). The wisdom derived from this type of approach is tacit and profound. It leads to experiences that attentional meditation alone cannot. Ultimately it is the insight from the meditational experience of Niroda that leads to liberation (Engler et al, 1986). Watching yourself appear and disappear into the unity of all things has a profound impact.

The neurologist, James Austin (1998), investigated Zen meditation from a neurophysiological perspective and came to similar conclusions. Although he had profound experiences during the meditational process, it was the sudden insight that later struck him, when he least expected, that most astounded him. In Zen, one is given a koan or problem to solve, as well as meditational practice. It is often in the struggle to use their attentional focusing skills from meditation practice on their koan that students find satori, or the great insight.

We should conclude from these explorations that alpha training by itself is not sufficient, but that insight exercises are also crucial. To this end, we present the clients own life dilemma to them as the koans which they must solve. We find that synchronicity intensifies in their life at this point and their daily life becomes a powerful reflector of the issues at hand. Dreams, symbols, and life events begin to mesh and integrate in a manner that defies explanation.

Anna Wise and Insight

In the 1970's, before the advent of quantitative EEG, a British investigator by the name of Maxwell Cade (1987) invented some specialized EEG equipment for meditational training as well as to record the EEG of accomplished masters of various religious traditions and other peak performers. One of his friends, Anna Wise (1997), took his findings and formalized it into a program of self-transformation. Others, such as Adam Crane and myself have arrived at a similar place through different but similar avenues.

Anna's approach is interesting because she does not rely as heavily on the feedback aspect of EEG technology for her program as many other in the field do. She tends to train students in techniques, often working in groups, and then monitors the results of each individual with the EEG. Many of these techniques are visualizations aimed at achieving specific brain states with specific EEG signatures as well as specific insights. She observes a unique pattern of brainwaves in individuals who attain high levels of insight and argues that this pattern comes to dominate an individuals mind if they are consistent enough in gaining those insights. We might call this achieving the insightful mind. Anna calls her technique neuromonitoring and the resulting outcome the Awakened Mind pattern. Insight then, alters the EEG patterns of the brain; eventually, in a permanent manner through constant practice.

Clients and Visualizations

As we have previously mentioned, Richard Davidson (2000) found that depressed individuals tended to have under activated left hemispheres. We often see this in the form of excessive amounts of low frequency brainwaves on the left side of the brain in our brainmaps. By this we mean high amplitude 8hz alpha dominating in that area (see chapter 6). Based on this research, a demonstration was arranged by a major researcher in the filed of neurofeedback, Elsa Baehr (1997), at one of Bob Gurnee's workshops in Phoenix Arizona in which we monitored the percent of alpha in each hemisphere as a grad student focused on depressing thoughts. As her face became strained and tearful, you could clearly see the shift in alpha to the left hemisphere as Davidson's research predicted.

I had been using a similar technique in my clinic based on this phenomena to train my clients who were depressed. It was inspired by some of the original clinical experiments by Peter Rosenfeld and Elsa Baher (1997). We called it bilateral alpha training because we trained alpha amplitude up in the right hemisphere while at the same time we trained it down in the left hemisphere. We did this by using a special piece of EEG equipment invented by Hershal Toomim (a pioneer in alpha training equipment) that allowed us to "ratio train." This means a reinforcement tone increased as the ratio of alpha on the right increased with respect to the left. We had dramatic results with this procedure on most of the clients we used it on. All of them had been chronically depressed and heavily medicated for years.

One of the clients we were using this technique on was having an especially difficult day, as was usual for her, shifting her alpha to the right. Consequently, I decided to use a brief inner child visualization that I often employ in my training. We did a pre and post visualization baseline and found a dramatic shift in her alpha amplitude from the left hemisphere to the right. Her training, for the rest of the session, was far improved and she attained measurable results that were dramatically better than anything she had done in the past. Obviously, we concluded, visualization is a very powerful tool for immediately altering neurochemistry and neocortical dynamics. The Tibetans have created very sophisticated visualization exercises to promote and develop this ability and utilize it to train their monks. It is clearly a valuable tool for self transcendence.

The Journey

I encourage my clients to look on their lives as a journey. Their journey is about their own unique pattern of growth into maturity. It is frequently a painful journey. It is not about acquisition of social resources, such as wealth and power (although that may be part of it), but it is about understanding fully their own potential as a human being. In western culture there appears to be stages extending from birth to death involved in this process. To a large extent, I believe the psychologist Erik Erickson presents the stages of that experience very well.

We are primarily social beings and we are born into and develop within a sort of social dialogue and common social dream. As we grow, our social boundaries expand. We become progressively more capable of navigating the various social worlds we encounter. We do this not only by becoming more knowledgeable about the different aspects of our society and how they work, but also because we grow in our skills of self-disclosure and intimacy. Productive outcomes of interaction, as Daniel Goleman (200) in his work on primal leadership has shown, is a consequence of emotional intelligence and the strong relationships it fosters. Whether we are in business, the arts, or politics, strong social relationships even in the domain of weak social ties results in positive growth for our group or organization. Within the family these same interpersonal skills contribute to harmonious relations and personal growth. Relationships deepen as our capacity to know and understand others increases. This capacity is born out of extensive experience that involves profound and ongoing self-examination, although many of us avoid this examination process. This growth is infinite and without end. At some point, however, we cross a threshold in which we come to fundamentally grasp the process. We come to see the universe as a means and not an end. It has no one overall meaning because it contains all meanings.

The Asymptotic Paradox of Rational Thought

Asymptote is an algebraic concept which is not difficult to grasp but which has profound implications with regard to neural networks. As one follows a line defined by an asymptotic equation, one finds that the destination where it should cross the axis of a Cartesian graph is never arrived at because the line curves slowly until it parallels the x or y axis. In a similar manner, when we seek the truth or the definition of a word, idea, or concept, the closer we get to it the more we lose its definition. As we isolate the identity of an object it evaporates because it cannot exist as a thing in itself.

The philosopher W. Quine (1953) gave us the profound insight that all concepts or cognitions are defined by all others. The richness of understanding of a concept depends on how rich our experience is of all factors that contribute to that concept. A concept is therefore not a thing in itself but a position of relationship in a network of other concepts, images, feelings, etc. This matches our present understanding of neurophysiology at present. Antonio Demasio in his book, The Feeling of What Happens, explains that memories, of which concepts are a subset, are reconstructions in real time of pieces of memory from different functional areas of the cortex. The image comes form the visual cortex, the sound from the audio cortex, movement from the sensory motor cortex, etc. They are brought together and integrated as an internal experience. A concept is therefore an internal experience. No thing "is" therefore of and by itself but rather a reflection of other things through relationship. All things are consequently a different and unique experience of one thing, the field of experience. Any given thing is merely a unique position in the field of all things. Nitche noted that beauty was but a component or reflection of perfection found in a particular object. To this end we could say that *truth is the perception of unity in any given field of experience.* Grasping the relation of an object to the whole, in all its details, is to understand the object (even the self) to its deepest extent. It is also to grasp the implicate whole as well.

Understanding at this level takes time. To explore an entire system of experience in relationship to one coordinate in that system could be an infinite process; but each iteration brings you closer to the experience of the whole. A lifetime in the human field of experience begins to prepare us to grasp the whole of that field of experience, infinite as it may be. We begin to grasp the perfection of our field of experience through particular experience and self-examination. That experience of perfection is the experience of the whole and our fundamental relationship to it as a part. We share in its perfection. With enough iterations through the network of experience we come to fundamentally grasp that perfection and it alters our experience in a profound way. It cannot be spoken or explained because it requires a lifetime to explain it. As Lao Tsu says "he who says does not know and he who knows does not say." We might amend that end to "can not say."

Our journey, then, is to arrive at this understanding through relationship in the social order. Our biology and its social manifestation are vehicles for integration with the infinite field of experience. We become perfect reflections of what is before us. As they say in Zen, we are polishing the bodhi mirror. This may all sound too abstract, but it is very concrete in experience.

Entering into human relationship from this perspective changes the meaning of the game entirely. We no longer are completely caught up in the socially defined goals, only caught up enough to make the process interesting and to play our roles properly. We are playing our part with our awareness always on the prize, the core experience of unity. Life tends to become more like play at this point. Communication seems to become less driven by the automatic side of ourselves because we are more fully present. We are less habitually dissociated into internal iterations of fear based networks. The amygdala does not have to work so hard. We are not constantly programming it through classical conditioning to be highly vigilant for dangerous stimuli. With less of an agenda the limbic system calms down and cooperates with the cortex better. We tend to depersonalize situations somewhat and those things that used to elicit intense negative responses seem muted. We spend more time in alpha and have more resources to fully focus our neural capacity on problems. Our arousal level becomes more even and we don't become as easily overwhelmed in situations which require complex cognitive functions. My clients rediscover and describe this component of their training experience to me constantly, although for most of them it is only superficially articulated and does not contain the wisdom component that comes from profound insight. A little nonattachment can go a long way.

All of this is not to say that my clients arrive at perfect insight in ten sessions. It is rather to demonstrate that this experience sets up initial conditions in a chaotic neocortical dynamical system which has alpha as an attractor state. Over time they begin to attain greater clarity and insight because the system is working better. Aversive external social conditions can short-circuit this process to a considerable degree in its early phases, but given enough time the process becomes robust enough to sustain itself through intense challenges. My clients begin to feel themselves slipping into anxiety or depression, but suddenly pull out of it. They recenter in their system's optimal coupling zone. It takes time for them to trust that they are not going to get stuck outside that zone anymore. As they become more aware of their dissociations they integrate them and let them go. They no longer perseverate on them. They stop getting "hung up" as we say in the vernacular. This process we are describing was first explored in some depth in the book Mindfitness by Adam Crane and myself.

Tax Collectors and Thieves

In this sense, the psychophysiological process is connected to the spiritual process. It is a journey into biopsychosocial maturity and wholeness. Wholeness is the underlying theme to the entire journey of recovery of mental order. To some degree I find that those who are most engaged in this process are those who are experiencing mental disorder. They are in crises and are motivated to confront their dissociations. They are terrified and resistant, but they realize that they are already walking through the fire. They are just not sure if there is another side to it all. This is the most frightening aspect for them.

Most addicts must first punch through their own denial before they can deal with their issues. The denial centers around the fact that they are suffering and using the drug of choice to cope with it. Pleasure is the western antidote to pain, not wisdom. Consequently we have built a society of distracting pleasures and delights. We have an abundant supply. They keep us distracted and safe from the pain. Soon we forget why we have engaged in the distractions, we just become desperate when they diminish and we come too close to the pain. Our problem becomes not one of dealing with suffering, but one of making sure we have a good supply of the "stuff" we need. The denial grows out of intent but feeds off of habit and forgetfulness. The lotus-eater forgets why and how he arrived at his destination. Unfortunately this denial does not become confronted until the addict "hits the wall" several times. When they are truly isolated, destitute, and alone, addicts begin to fully experience their suffering and open up to help. This is often true of other disorders. Individuals hide their problems year after year until they become out of hand. So it is almost axiomatic that those most ready to receive guidance are those who are experiencing the poverty which the side effects of mental disorder generate. This disorder cannot be separated from social reality. It is born out of social contexts and comes to live in the biology. It is no mistake that Jesus sought to teach among the poor and discarded populations.

8

Faces of Confusion

*The definition of insanity is that we keep doing
the same thing even though it doesn't work.
The Big Book*

The Journey I am describing involves not only psychological integration but also the generation of greater order into the most complex system we know of in the physical universe, the brain. The brain achieves its full physical development between the ages of twenty-one and twenty five. This is just one step in its evolution however. The next decades will involve ongoing integration and consolidation. Organization continues on a network level and is an ongoing process whereby dendrites grow and dissipate as new neuronal networks engage and disengage each other. Receptor populations in the neuromodulator systems increase and decrease as vigilance, memory, and emotional networks wrestle with changing cognitive valences relating to social interaction and the associated social reality. All of this activity is then in response to the social challenge.

We can document the shifts in these systems in terms of neocortical dynamics to some degree through the science of neurometrics. We can measure and topographically map the standing EEG wave patterns which emerge from the constantly shifting electrical fields generated by neuronal activity. Frank Duffy (1994), E. Roy John and John Hughes (1999), Bob Thatcher (1998) and many others have worked to develop a normative EEG database. They have found that there is a relatively normal distribution of the EEG across measurement domains of EEG power, magnitude, coherence, phase and frequency. Of course, the concept of what constitutes a normal personal is highly debatable, however, the standard used for this particular situation was a variety of psychological tests that excluded individuals with extremely deviant behaviors. This provides a starting point for analysis. It goes without saying that all of us involved in this process would like to uncover a definition of supranormal and transcendent. Richard Davidson (Goleman, 2003) is exploring these dimensions when he measures altered states achieved by Tibetan Buddhist monks. His findings continue to increase our knowledge of the unique states of consciousness associated with meditation. They suggest to us new ways of understanding and using the brain as well as how it relates to the concept of mind.

As a result of these emerging imaging sciences we have come to understand that there is a fairly normal distribution of activity in the brain which is to be expected from the average Joe-or Jane. This activity is a reflection of a healthy level of integration and order in the brain that allows individuals to participate in a positive manner in the social order. We can also observe from these types of imaging when the brain is showing a lack of order and not functioning well. This lack of order correlates with severe suffering and behavior that is socially inappropriate, destructive, and often self-destructive. We can even use these techniques to observe where and how the brain is disordered. We know a disordered brain and a disordered mind are highly correlated. Presently, we are beginning to understand how a super normal brain might function by imaging experts in concentration and meditation as well as other peak performers in more socially interactive areas of life such as business and athletics.

Waking up involves utilizing concentration skills, visualization, and insight to disrupt the habitual occupation of endless cycles of automatic trances we pass through everyday. These trances have a powerful emotional attraction that pulls us into them and they are grounded in our childhood socialization process. As we wake up we begin to experience the emptiness of our pursuits and begin to see them as either running away from pain or running into pleasure in order to avoid the restlessness and agitation that emerges from daily life. This daily distress in turn is largely a result of the way we frame our moment to moment reality. This framing shapes the way we interact and negotiate our social and physical environment. Our environment and bodies begin to encode and embody the patterns of our inner life, reflecting them back to us. These more physical factors exert an inertial force that continues to draw us into our trance states and automatic reactive patterns. It is a powerful feedback cycle. Liberating ourselves requires vision, passion, dedication, guidance, self-discipline, self-awareness and relentless effort. It is a powerful inertia we must overcome. It is our own socialization process.

In the process of waking up, our insight, which guides us to our wisdom, provides key schemas or mind sets that allow us to organize our new emerging experience of life in a more integrated and harmonious pattern. This transformational process becomes the platform for the leap into direct experience of unity.

What follows are some core schemas for re-organizing experience at a new level. It requires the motivation and self-discipline to continually observe and inspect your own behavior. Through this effort at re-organization the individual becomes increasingly aware of his or her self-process. This process can then be slowly reshaped so that the self becomes more integrated and harmonious in its function. As this occurs powerful insights lead to progressively more intense transcendent insights. With the additional tools of increased concentration, visualization, and compassion from the meditative training, the individual is able to very powerfully transform the self into a more transparent entity. This transparency allows the more universal aspects of self and awareness to emerge from the self. This new emerging more generic and universal self is more capable of experiencing unity.

Concentration allows the individual to powerfully focus on issues as they emerge and penetrate their meaning. It also provides increased self-awareness. Visualization provides the ability to more effectively and consciously program our behavior in the future, allowing us to adjust our behavior based on our awareness and insight. Compassion exercises integrate cognition and emotion, provide more emotional stability, and promotes integration as well as more effective interactions with others grounded in emotional intelligence.

The fuel for this transformative process is everyday experience emerging from the lifeworld of the individual. Negative emotion has a powerful ability to make us self-aware. When we are in situations that generate physical and emotional discomfort we have the opportunity to awaken from our round of trances. The brain naturally does this based on survival mechanisms. Everyday life generates emotional friction for us because it often deviates from our expectations based on our cognitive emotional map of everyday experience. We find these deviations threatening and they challenge us to work harder. We say we have encountered a problem and we begin to think and worry over it until we find a resolution. This provides the opportunity and the impetus for us to transform if we have the right schemas available. We can take the friction and the situations of daily life and use them to awaken and transform the self. The schemas provide a program guide for that process. What follows are some of the basic schemas you can begin to work with to generate positive transformation.

The Purpose of Daily Life Is Transformation

We have many goals and objectives that we persue in order to get our work done and meet our personal needs. These are important tasks. As we engage in them day after day, hour after hour, they become a sort of mantra that entrances us. We become so emotionally involved with these goals that our vision narrows and they become all encompassing. We lose our perspective. One of the first steps in transformation is grounding our awareness in our fundamental goal, transformation. To do this we must daily remind ourselves that this is our purpose. The primary meaning of the events of our daily life should be considered as fuel for learning, insight, and transformation. When we experience friction, pain or discomfort we should remind ourselves that the primary meaning of this event is that it will become an opportunity for growth and transformation. This can help us to pull back a bit emotionally and work more powerfully with the situation. Children are overwhelmed by the emotion of the moment, adults develop the inhibitory networks to endure the first rush of emotion and quickly recover their composure to work with the situation.

Managing Emotion

Emotions are to be managed not repressed. Research indicates that without emotion we cannot make good decisions (Demasio, 1994). Emotions help us discriminate between ideas and objects. People with good emotional self-awareness are able to make decisions easier and more accurately. The work of Daniel Goleman (1995) on emotional intelligence has demonstrated its enormous value to a culture that long ignored it. For the behavioral neuroscientist today the question is what is the best way to manage emotion? Once a strong emotion appears we are in its grip, we are destined to play it out to the end. A basic strong healthy emotion can last about twenty minutes. A pathological anger or sadness can go on for days, weeks, even years. On the other hand, one strong emotion such as despair can be quickly eclipsed by another such as joy. Emotional flexibility is a reflection of neural plasticity. It allows us to move rapidly from emotion to emotion. This openness to shift creates an opportunity for a constant background emotion of joy or bliss to move to the foreground quickly after each more temporal emotion emerges. But who feels a constant background such as joy or bliss? Meditators.

Meditation results in a unique subset of emotions we rarely engage in when we do not meditate. Bliss is one such emotion. It is almost like the emotional systems version of muscle tone. When we exercise the brain through concentration it results in a sort of flexibility or plasticity and what Adam Crane and I refer to as Mindfitness. The well-toned emotional system generates a constant background hum of bliss. It is very subtle and yet its constant presence makes it very powerful-like gravity. It is a constant companion and reminder of situation. Doing meditation is like practicing Judo. We train to deal with our more negative emotions because they come suddenly, powerfully and they are automatic. Because of our constant awareness we may not be pulled into negative emotions that arise out of excessive attachment to thoughts or objects. If we are pulled in, we flow with the emotion and roll quickly back into our background bliss. We are in balance and have a flexible stance. We experience our emotions, but they are at arms length and we recover very quickly.

If we try to ignore or suppress our negative emotions we are really ignoring our boundaries and our attachments. Learning, growth, and self-respect cannot emerge from this stance. It is a rigid stance. Anger is a good emotion when it is experienced at the appropriate level because it protects us. We have an obligation to respect and care for our bodies and our self-system. They are vehicles for navigating this world and learning. It can also alert us to attachments and agendas we did not realize we had. There are many traditional systems that are afraid of the negative emotions and shun them. As modern scientists we can see how self-destructive this stance can be as well as damaging to others. If we do not offer others resistance when they violate us, then we are depriving them of the opportunity to learn and encouraging ignorance as well as more violence. As our wisdom and compassion grows, we learn to compromise ourselves with wisdom so that we can provide the space for others to grow and learn. We are able to apply our inner emotional Judo so that it becomes an external emotional Judo. When wisdom and compassion encounter conflict, they transform it into learning and growth. Positive transformation occurs. Pain and suffering may occur as well, but they are also transformed. This is one of the hallmarks of true socio-emotional maturity.

Recovering Personal Power

As we become more self-aware we often find we have given ourselves over to many dubious causes. We have developed agendas and attachments to a long list of persons, places and things. To maintain these agendas we have made ourselves subject to the decisions of others to a very high degree. Our emotional slavery can inspire us to threaten and cajole others into maintaining the course we want based on our other agendas, so we become petty tyrants. In this chain of fools, one person's slavery becomes another person's tyranny. A contest of automatic agendas surfaces. Frequently, addiction to the attention, love and approval of one person can buy us an immediate subscription to this chain gang. In many immature romantic relationships, individuals will sacrifice their entire emotional awareness as well as physical domain to feed this addiction. Often this can solidify into a marriage that sours with silent anger, resentment and depressions.

Drawing boundaries is an essential solution to one aspect of this problem. Frequently we find that many problems issue from the fact that clients have developed a life pattern of doing what others want in order to gain their love and/or approval. This has won the term "co-dependence" in counseling circles. We say they live in the face of others. That means they go through their day carefully watching everybody's face to make sure no signs of disapproval appear. They constantly and unconsciously morph themselves to avoid disapproval. If they do encounter it, either they panic and feel crushed or they retaliate in anger. This is a reactive stance toward life. An individuals decision making power is sacrificed. Like a leaf in a tempest there is little time for self-awareness or deliberate decision. Anger, resentment, and negative emotions thrive.

The first step into recovery is to become aware of your bondage. The second step is to start using the word "no" on a regular basis. Learning to accept your role as "bad" guy in someone else's melodrama can be difficult. Many of the relationships you have forged through servitude may diminish with time. Discovering and believing in your own fundamental value as a human being can also be a challenge. Establishing relationships with individuals who know how to engage in mutually respectful exchanges can be a revelation. It is also a surprise how quickly anger, resentment and depression disappear.

Speaking Your Truth

Leaning to express your feelings, instead of telling others how to be, through specific strategies of defense is a difficult process. We learn at an early age that if we reveal our emotions to others they may use that information against us. They will use the information as evidence to support their position of persuasion. They may make fun of us, minimize the importance of our feelings, or tell us our feelings are invalid. To protect ourselves we often hide our feelings and motives and use a strategy to get what we want. In many cases we are hiding our real fears and making demands or arguments to get what we want in order to protect ourselves and maintain our agenda. We might be concerned there isn't enough money in the budget for a vacation but instead of revealing our concern we argue we should save up for a bigger one next year or that we should use the money for something else. We usually engage in this behavior automatically and with little awareness. It usually results in endless arguments that never resolve anything and confuses others because no solutions ever seem to emerge. This habitual type of behavior also leads to an inability to understand our own motives and prevents us from ever resolving many of our real needs.

Engaging in this behavior can escalate in relationships where others are power seeking, manipulative, controlling, demanding, insensitive and out of touch with their own emotions. If they fail to listen to us, honor our emotions and compromise, then we may naturally begin to hide our feelings and attempt to get our needs met through other means. Deception, argument, and avoidance are possible strategies that may result. Allowing ourselves to enter into such a relationship may grow out of past experience with similar relationships, especially dependent relationships such as the one we had with our parents.

Self-awareness is an important key to the process of dealing with automatic camouflaging of our motives. We encourage people to use "I" messages a lot. Telling other people how you feel gives them more specific information about how to respond to you without violating your boundaries. This is particularly important in intimate relationships. Using phrases like "I am concerned about…" in more formal relationships lets others know the real issue and this brings focus and clarity to the discussions. Using a phrase such as "I am afraid that…" or "my worry is that…" in more intimate relationships tends to short circuit older more automatic sequences of behavior. These statements also promote trust and honesty in relationships. However, they are of limited value if the other individual does not respond to them with the same directness, honesty, and trust.

The Importance of Being Right

Being right comes in two forms: being right by making others wrong, and being right by making yourself not wrong. Either one is an expression of low self-esteem and immaturity. It is a key feature of childhood wounding from over controlling parents and siblings-usually with the same issue. Being wrong comes to imply weakness and defectiveness. People become embarrassed and ashamed of their mistakes. They hide them and pretend they did not occur. They may point the finger at others in order to keep their own mistakes a secret. If their mistake begins to emerge they may become defensive and angry. They may take an offensive stance and begin recounting mistakes others have made. Many of us develop a negative attitude toward mistakes in school where the wrong answer can result in classroom humiliation. Growing up in a dangerous environment can mean a mistake could be extremely painful or fatal. This too can condition individuals to be extremely fearful of making a mistake.

The problem with this response to error is that it makes learning and growth difficult. Making mistakes and correcting them is a natural part of learning. If we have selective areas, particularly in the realm of human relations, where we are afraid to make mistakes, then it will be very difficult to correct our errors and grow and develop. It prevents us from developing openness and humility. It becomes a major communication block and prevents us from solving problems with others and developing realistic compromises.

Doing It For You

Clients begin to see the value of confronting others and making statements of feelings not because it makes them right or straightens the other person out, but because it makes them more self-aware. They become acutely aware of what they feel about things. With work, this can become a moment to moment awareness of what they are feeling emotionally and thinking mentally. It is a form of honesty that enhances growing self-awareness. Interacting socially based on this clear self-awareness improves relationships with others, makes your behaviors more congruent with your goals and intentions, and results in outcomes that move you forward into positive directions of growth and learning. It provides opportunities to explore new areas, make mistakes and grow in wisdom as a result of the process. Your behavior becomes centered positively around your self-awareness and not reactively around other-awareness. This latter approach results in masking behavior, hiding, avoiding, resentment, anger, manipulation and multiple violations of self without conscious acknowledgement.

Recovering Memories

Often we don't lose traumatic childhood memories; we just define an event as something else and store it under that name. Sometimes we have defining moments when an event dramatically changes the way we think, but we become so engaged in our new agenda that we forget what precipitated it. When we begin to carefully review our memories with new eyes, we see them for what they truly are to us. We see them with the eyes of an adult instead of a child. Often as children we file things under headings which make sense then, but from an adult perspective look completely different.

The conclusions we draw about life from these memories or core events give us a general orientation in life that results in a habitual way of thinking, feeling, and doing as we react to unfolding life events. We develop characteristic ways of behaving in various situations that others see as patterns of personality. We may be seen as a "go getter" or "picky" or having a "negative attitude." Others tend to cast us in roles in their personal life dramas and we tend to do the same with them. Out of our key life events we draw conclusions and build agendas that result in an ongoing life drama or core drama. We may have one core drama or several overlapping ones going on at the same time.

Core Dramas

For many of us there is often not just one core event but several smaller events that work synergistically together to define our perspective and agenda in life. This agenda defines our expectations for the future and what we expect to get from others when we interact with them. From this arises our core drama. We usually play these dramas out without being aware of what we are doing. Like fish in water we swim through them from day to day unaware of the medium we are operating in. It is very difficult to become aware of your core drama because it is so fundamental and habitual that you are "it." Once we do become aware of these core dramas, powerful insights unfold which can transform and liberate us from our sleep walk.

One of my clients was raised in a family of several other sisters. As a younger sister she felt she could not compete successfully against her other sisters for her parents attention and approval. The others were cast in the family drama as smarter or prettier or more athletic. Since everyone was so busy and the house was such a mess, she found she could get attention and approval for cleaning up and helping out with the details of running the house. She became the responsible daughter. This was her core drama.

As she grew up and began to work she would play this role at every job. She worked doggedly and put in extra time. She was quickly promoted to managerial positions. Others became incompetent and irresponsible in her eyes. She was always picking up after everyone and having to do their part. As a result she began to become constantly angry and frustrated, then depressed. Having taken on so many other peoples details to make her department look the best she was becoming resentful and burned out. She never confronted others and so they continued to let her take care of the details. If she did confront them and they started doing a better job, then she would lose her opportunity, her stage, for her performance. She could no longer be the "only really responsible one" at the office. She would lose her "leg up" on everyone around her. She was still acting out her family drama. What is worse, the same drama was unfolding at home. She was exasperated by her lazy husband and irresponsible daughter. Her daughter never did things for herself and was constantly in trouble. This client became severely depressed and developed migraines and fibromyalgia.

Another client came to our office with severe panic attacks daily. As we worked with her and began to calm down her central nervous system memories began to emerge. She had been overwhelmed by her work and her responsibilities at home. Her son was having problems with a mental disorder and she felt it was because she wasn't giving him enough attention. Her house was a mess and she couldn't concentrate at work. She felt like a failure. One day while she was training with neurofeedback she let out a gasp. She said "I *don't* have to do everything perfectly, do I." I replied, "of course not." She then explained that her father was always criticizing her and correcting her in everything she did. As an adult she would spend hours at home making sure everything was perfectly clean and in its place. She realized that she was such a perfectionist about everything that she could no longer get anything done. Once she had this insight, she was able to begin letting go of the details and her panic attacks subsided.

Of course this did not happen overnight and it required the assistance of neurofeedback training to renormalize her nervous system, but it did happen over a period of a few months after she left our office. We called her two years later and she was still medication free and had no more panic attacks.

Parenting Yourself: Firing Your Parents

It is important to become aware of the childlike pattern of emotions that often direct our actions and take better control by self-parenting. Since the majority of our basic emotional scripts are learned in the first two years of life before we are even aware that we are a separate self, they are deeply embedded in our processing below our daily level of self-awareness. They continue to be built upon and actively shaped through our interaction with our family and friends. We usually react emotionally and automatically with little self-awareness. Thus, most of our responses are shaped and determined by the past. Once released, we have little recourse but to play them out or try to hold them in. Hiding them usually leads to mental disorder.

In a sense then, we continue to respond to ourselves and others much in the same way our parents did. Much of our reality is shaped by the cultural and familial baggage they carried. Some of this material may be inefficient, self-destructive, or harmful to others. In order to transform ourselves beyond these limiting beliefs and behaviors we need to see them, acknowledge them, and see their limiting influence on us in terms of our potential transformation. In a sense it means we need to fire our parents and take over the job ourselves.

The next step is to substitute new behaviors that are more efficient and lead to liberation from our automatic reactive self. To do this we must employ self-awareness as much as possible at all times. Like a mother with a two year-old we must be constantly vigilant and watch our every move. When we make the wrong response, we must stop ourselves and inhibit it while replacing it with a new response. For instance, if we are walking along and suddenly find ourselves in the midst of a negative emotion, we need to stop and recall what thought triggered it. We then need to evaluate the thought. Usually we find it is an intrusive thought that is automatic and not well grounded in reality. Such automatic negative thoughts can then be noted and watched for in such future situations. When we catch them as they emerge we can stop them and replace them with a more realistic response. This tends to mitigate the attending negative emotions. One has to be very aware and very fast to do this, but it can be done with practice.

It is important to note that when the negative emotions emerge, there is nothing we can do about them except watch them play themselves out. It is also important, however, not to let them spawn new negative thoughts which we become involved in. Any new negative thoughts should be recognized as such and politely told to go away. The negative emotion will usually die away on its own in 20 minutes or so. Anything longer indicates a bigger older issue that is being fed by other automatic thoughts we have yet to become aware of.

Fatal Distractions: The Ways We Get Lost

Another whole dimension of factors that keeps us at a level of low self-awareness and automatic behavior is the pursuit of pleasure. So far we have clearly outlined the factors based in fear that defined our lives in terms of fearful reactive behavior, but we have paid little attention to what often draws us forward and propels us into new directions where we become lost in what we want. We have both negative and positive attachments.

The negative and agitated state we are frequently in due to automatic thoughts and feelings regarding our fears can also be relieved through distractions. Exciting distractions inhibit our awareness of other stimuli both internal and external. The type of distractions we choose can be as novel and specific as we are unique. They often involve formal forms of culturally recognized recreation such as sports, going to the movies, or concerts. They could involve gardening, shopping, eating, driving, reading detective stories or emailing friends. The list could be as endless as the types of things we worry about.

Some of these distractions can be negative in nature. In fact a good horror movie or novel is, at present, a very popular form of distraction in our culture. This type of entertainment creates such a powerful experience of fear and excitement that it wipes out the sensation of any other feeling that might be present. Any other forms of extreme entertainment tend to do the same thing. Extreme sports such as skydiving can wipe out negative psychophysiological backgrounds for days. It is well known that skydivers get a high from the sport that lasts for days afterwards. This is an extended form of parasympathetic rebound in response to surviving the extreme sympathetic arousal from confronting death so directly and immediately. It is a form of mild elation.

The problem with these extreme forms of distraction is that they are powerful mood enhancers, which like drugs can become addictive and demand increasingly higher doses of intensity. Now we have many forms of extreme skydiving. Our technology continues to develop and provide us with new innovative or extreme forms of entertainment. This is fun, to some extent, but it is only a temporary solution to our negative background mood and agitation. The only real solution to negative mood and agitation is to directly confront it, uncover its source, and consequently remove it. This of course goes beyond transformation and moves us into transcendence. It is important to observe and acknowledge in your own behavior that not only do you avoid things in life, i.e. run away, but also that you pursue other things to try and avoid those same things, i.e. run toward a solution or relief-even though it is only temporary. Pleasure derived from recreation and distractions is usually an escape from discomfort.

The problem with agitation, restlessness and discomfort is that we have all experienced it for so long in our lives that we have habituated to it. We don't notice it in the background even though it has been driving our behavior for most of our lives. It is insidious because it appears to be mild in comparison to other forms of pain and discomfort, like boredom. Yet it exerts a powerful influence on our lives. I have seen many children cry because they were bored and even go into fits of anger and rage. They require a certain level of stimulation or over stimulation. Adults just have a more sophisticated form of this as well as elegant ways to deny it and hide it from themselves. As an experiment try just sitting quietly and do nothing for an hour. If you are well rested fed and comfortable this should be no problem. A majority of adults, however, find this a very difficult task. Why is that? There is an underlying restlessness and agitation that usually goes unnoticed in the background until they are in a low stimulus environment. In response to this situation, many people will fall asleep even though they are not tired. A famous Russian scientist, Ivan Pavlov, working with the classical conditioning of dogs, noticed they had the same response to the internal tension of a frustrating situation they could not resolve. It may be that for human beings this situation is the dilemma of life itself.

The Neurofeedback Side of the Training

When Adam and I wrote Mindfitness it was very well received by professionals but some others were looking for more of a cook book on the topic. Although there is a great deal of theory in this book that should provide insight, there is also the ISI which follows to provide a concrete outline of where in your lifeworld you could benefit from change. Once you take the ISI you, will need to fax it or send it to the address or number at the Mindcycles.com website so it can be scored and interpreted. In the near future you will be able take the ISI online at the website. Many people also find they want some coaching with it.

Experiencing neurofeedback is another issue. If you want to try it, you really need to contact a neurofeedback practitioner. You can locate one near you through listings at www.bcia.org or www.eegdirectory.com. It is not initially a do it yourself technology. The equipment is also FDA regulated. Neurofeedback practitioners are individuals who have an educational background in health or psychology and go through a long training period to become certified. Once you are established in your neurofeedback training, many practitioners will allow you to do home training. If the practitioner you work with does not do home training, then you can contact us at Mindcycles and we will get you started if you qualify. If you like, we can help you with the whole process from A to Z. Just email us through the website.

9

The Interactive Self Evaluation Instrument

*We don't see things as they are,
We see things as we are.
Anais Nin*

The Interactive Self Inventory was developed to assist fairly normal individuals identify areas of their behavior that were generating problems with respect to transformation and integration. A basic level of self development and integration is necessary in order to achieve access state. As we have discussed, an individual with unresolved issues leading to inappropriate social interaction and poor social accuracy results in a life world full of intense conflict and upheaval. The level of anxiety, distraction, and agitation may be so extreme that it leads to episodes of depression. An individual in this situation will find it all but impossible to sit and focus appropriately on the issue of transcendence. For the majority of westerners this appears to be the case. They handle transformation, although with considerable difficulty, much better.

Using this test as a guide to areas of difficulty, individuals can engage a coach or counselor of their choice to work through these issues in preparation for training in access state. The two activities do not have to be mutually exclusive and can both be engaged at the same time. In fact, they do appear to be synergistic in their effect. Individuals should not, however, expect great strides in transcendence until they have resolved their issues of transformation. They will likely find that training in access state facilitates the process of transformation. We have found this to be the case over the last decade working with neurofeedback as an adjunct to counseling.

What follows is a discussion of the rational behind the instrument and its dimensions as well as a discussion regarding its structure and how to interpret it. To score the test will require contacting the related website mindcycles.com.

Unfortunately this is necessary at this time until the inventory is more widely distributed among individuals who have been trained in how to interpret it properly and advise individuals accordingly. In the meantime you will receive a basic response with general information for how to proceed based on the results of the evaluation. It is imperative to find a professional coach or counselor to help you work with the results, or you can contact one of our trained staff to assist you. Just doing the test alone and reading the results is not enough. It is only useful if you act upon it appropriately.

Using The ISI

There are literally thousands of instruments for evaluating individuals and they all tap different and unique dimensions. Human beings are so complex however that we could keep inventing new instruments and measuring them forever without exhausting all possibilities. The problem with early efforts at personality and trait assessment instruments was a lack of underlying theory regarding the self as a process. What is needed to develop a really good instrument for global personal evaluation is a sound theory for identifying and articulating the basic important dimensions of behavior.

The basic concept around which the inventory is constructed is Social Accuracy. This concept links biological, psychological and social theories together and draws upon the related disciplines to construct a more complete and holistic concept of self. The basis of a positive and successful *self*-experience is that it results in effective social interaction resulting in the achievement of personal goals, personal growth, and personal health.

Social Accuracy is the ability to navigate the social system in such a manner as to derive the resources you need to grow and thrive without undue harm to others. It results from dimensions of cognitive functioning, emotional processing, and self-awareness. The appropriate development in each of these dimensions at each stage in the growth cycle results in effective action that secures the individual the resources he or she needs. It is the key dimension to personal transformation.

The end result of Social Accuracy is maturation. As a consequence of years of interaction between the self and the social order, the self matures and integrates accumulated experience. This maturation process results from insight leading to accumulated wisdom. Wisdom is a higher form of understanding than intellectual understanding and it involves cognitive, emotional, spiritual, and biological dimensions of understanding or knowing. It is the integration of experience and analysis. Once sufficient wisdom is accumulated, then the personal transformation cycle diminishes as the personal transcendence cycle initiates.

Research to date suggests that an individual cannot really engage the transcendent cycle until the transformational cycle is relatively complete. As Richard Alpert noted, you must get a self before you have a self to transcend. This self, paradoxically, will generate considerable distraction and suffering preventing an individual from effectively pursuing transcendence until it is well integrated. Getting lost and making mistakes, it must be acknowledged, is a part of the growth and learning aspects of the transformational cycle. However, one can get lost and make mistakes to their own detriment; to the point where it is counterproductive even from a long-term growth perspective.

The dimensions chosen for the Self Evaluation instrument are derived from years of extensive analysis and clinical experience both in the areas of mental disorder and peak performance training in businessmen and athletes. This has been integrated with an exhaustive analysis of traditional modes of transformation (religions and metaphysical systems). The dimensions describe the modes of interaction that generate suffering and distraction. By identifying them in yourself and determining the level at which you engage in each mode of interaction, you can determine an appropriate behavioral change to make in order to reduce suffering and distraction and enhance Social Accuracy. Once again, Social Accuracy leads to positive outcomes and enhanced growth leading to maturity of the self. Strategies for change are provided. This process also contributes significantly to the development of self-awareness and emotional intelligence, which in turn enhances learning.

Method of Change

Change is very difficult. As Piaget noted, it involves restructuring of the entire self. Most people would rather ignore new challenging information than accommodate it, since often this would involve global re-evaluation. For us, the re-evaluation process involves actually trying to change behavior and evaluate the results in term of improved social accuracy. Since past behavior is habitual, routinized and therefore automatic, it is extremely difficult to observe and then change.

Once you have become aware of dimensions of self (through the Self Evaluation instrument) that can cause friction in your daily social interactions, you can begin to look for friction in your daily life that matches. One observes the emotions and thoughts related to the situation that results in friction, determines the definition of the situation and then decides how it matches the dimensions of self that are possible sources of friction. Doing this several times increases self-awareness in this area and eventually results in an insight into self-behavior.

At this point you would look at strategies for change and choose one to employ. Then you visualize a similar situation in which you employ the specific strategy for change and envision a positive outcome. Finally you wait for a similar situation to arise and then watch to see if you are able to employ the new strategy. You may have to try several times before you are successful. Patience is important.

In the course of doing this type of exercise, part of the insight you have may be related to Original Motive. Original Motive is a basic childhood belief that has never been consciously reviewed by your adult self. These motives usually are behind behaviors that cause friction in our lives. These motives stem from Primary Beliefs that we formed with our childhood mind. As we developed our self in the socialization process we observed behaviors of others and drew conclusions based on limited understanding of the definition of the situation. Often these were very emotionally charged situations. These Primary Beliefs became the basis for enduring motives that result in dysfunctional behaviors. These behaviors become automatic and routinized as they are based on fear and the drive for survival as it manifests in the protection of the self. We lose consciousness of these automatic behaviors and their motives and they emerge spontaneously in certain situations containing key stimuli that set them off. Another aspect of uncovery work is finding Original Feeling Scripts. Many of the feelings we have were learned before we were self-aware. We learned them through operant conditioning and vicarious learning watching others. As young babies our nervous systems tracked those of our caretakers and we learned emotional sequences automatically. These emotional sequences or Emotional Scripts emerge as we develop our self as a basis for how we should feel about things. Often as adults we will have feelings that are irrational and this is because they are based in these Original Feeling Scripts. These feelings can be changed through self-awareness and insight.

Another method of change is becoming aware of automatic thoughts. Automatic thoughts are constantly running in the background networks of our brain even though we are not usually aware of them. Two approaches are used to deal with them. One is Instant Recall. This is a technique where you stop whatever you are doing when you notice a negative feeling occur. You then think back in time until you can uncover the thought that triggered that feeling. That is an automatic thought. Automatic thoughts are usually related to an Original Motive. The other technique is self-awareness Concentration Training. During this training you watch quietly as thoughts emerge and practice letting go of them. We usually do this in conjunction with neurofeedback training.

The following material actually reviews each ISI dimension in detail. After you take the ISI and get it scored, you can look at each dimension and how high or low you scored in it. Using this information you can get a fairly objective picture of where you need to alter your thinking, feeling, and doing as well as you beliefs. This process, of course, works much better when you have a professional third party working with you while you are doing neurofeedback training as well. However, if you are an accomplished meditator, the neurofeedback is not really necessary.

The Dimensions of The ISI

Avoidant   Meta Category   Slow Left Hemisphere/Fast Right

Individuals high in avoidance are typically anxious in nature and frequently susceptible to depression. They are usually shy as children. It is common for them to be uncomfortable in social situations. Their self-confidence is low and they may be suffering from low self-esteem. They may see the world as dangerously unpredictable. They do not feel they have much control over things and are hesitant to act. They are reluctant to express their thoughts and feelings for fear of being ridiculed or manipulated. They have rigid agendas and dislike it when others disagree with them. They tend to entertain a lot of negative automatic thoughts and emotions (rumination & worry). They see the glass as half empty. They are at risk for being reactive and defensive. They tend to recover slowly from emotional upsets and negative events. They avoid conflict and confrontation. They often feel numb and have a muted experience of their own feelings. The may stuff their feelings and hide their anger. They often have trouble saying "no." They have poor focus, memory, and concentration.

Sub-Dimensions: Dependent, passive, inhibited, competitive, perfectionistic, impulsive

Interactive   Meta Category   Left Hemisphere Fast/Right Slow

Interactive people are self-confident. They are socially accurate in most situations. They enjoy novelty and see the world as an adventure. They trust their ability to respond effectively to others and are slow to judge them. They are usually high in emotional intelligence and can easily put themselves in the shoes of others. They tend to engage in mostly positive emotions and recover quickly from negative events and emotional upsets. They see the glass half full. They have good focus, memory and concentration. They are not afraid of conflict or arguments. They feel free to get angry and express their displeasure but they have good control over their feelings; they rarely let them overwhelm them. They are flexible, work creatively with situations and roll with the punches.

Sub-Dimensions: Independent, co-operative, flexible, assertive, regulated, relaxed.

Dependent

Dependent people tend to wait for others to take action. They like others to take the drivers seat. This way they can avoid being responsible for bad decisions. They tend to cultivate exclusive relationships and constantly seek attention from that relationship. They look to others to define the situation and base their emotional state on those around them. They dislike being alone and live in fear of being abandoned. Because disapproval is devastating they will avoid expressing themselves and tend to ignore their own thoughts and feelings when they conflict with others. They like to be taken care of and would prefer to sit back and let others make decisions for them. They let others draw their boundaries for them.

Independence

Independence can be healthy or extreme. At its extreme it is avoidant and controlling behavior that isolates an individual from others. It can result as a fear of being judged or controlled by others (inverse control). It is passively assertive because it avoids co-operation and consensus for the definition of a social situation. It may be an extreme expression of perfectionism wherein the individual can avoid the messy details of having to compromise. It can be a justification for an inhibitive style of being.

At its best it is an expression of healthy boundaries. Independent people are creative and enjoy exploring new ground. They often know how to sustain themselves without attention from others and have a healthy level of self-esteem. They thrive on adventure. They often make journeys into the unknown and return with gifts and discoveries for those that support them. They are willing to take calculated risks and will often push the limits of a situation successfully. They are not afraid of failure. They are not ruled by the feelings and ideas of others. They like to make up their own decisions and draw their own boundaries. They prefer to solve their problems on their own. Healthy independence reflects an awareness of the need for basic support and co-operation and will respect the rules of any organization, including the family.

Associated Dimensions: Assertive, flexible, regulated, perfectionistic, impulsive(?)

Competitive

Competitive people are group oriented, but for self-centered reasons. They want to access group resources in order to get attention. They thrive on attention. In a sense they get their attention actively as opposed to dependent individuals who get their attention passively. Their self-esteem is dependent on proving themselves better than others. They can often be very argumentative and they don't back down from confrontations. They may even cultivate confrontation and become experts in winning the argument. They give up easily and fear being abandoned by others if they are seen to make mistakes or fail at a project. They see mistakes as being deadly; an opportunity for your adversaries to take advantage of you. Competitive people are very vigilant of the actions of others and look for opportunities to denigrate them for their mistakes and gain the upper hand. They fear exposing themselves to the observation of others in novel situations and they like to maintain the upper ground where they know they are familiar with the risks and benefits. They are very rule oriented and know the rules of an organization very well. This allows them to manipulate situations to their advantage.

A mature sense of competition involves a lack of attachment to the outcome. It is often for the fun and excitement of the game and not to feed low self-esteem. It is subservient to higher motives and operates at the service of others and a love of the action of the game itself. The play is the thing.

Co-operative

Recent studies in many fields indicate co-operation is an important key to success and happiness. Two heads are better than one, whether its an emergency situation or a personal goal. People who thrive in a co-operative environment are often more mature and have greater emotional intelligence than others. They tend to listen more carefully to the ideas of others and let the group decide the best direction for itself or for the organization. Co-operative people know how to defer short-term gratification for long-term rewards. They tend to be better self-regulated and more in control of their emotions. They tend to trust others more and are willing to share their ideas and experiences openly. They tend, at their best, to be flexible and open to novelty. They are highly ethical yet patient with the mistakes of others. They are supportive and nurturing. They are not afraid of conflict or arguments if the group welfare is at risk. They tend to be more conservative in action.

There is a dark side to co-operation. When large groups of dependent individuals collect behind a highly assertive and independent leader, Groupthink can emerge. Dependent people will operate within the group for selfish reasons and compete for the attention of the leader. They may dislike the others in the organization or their ideas and suppress their own opinion while fostering resentment and anger among others. They may resent the rules of the organization and sabotage them. They may passively create problems and come to the rescue with a solution that wins them praise and attention.

Associated Dimensions: Flexible, assertive, relaxed, regulated

Perfectionistic

These individuals tend to always feel incomplete. They are restlessly engaged in always moping up the details and getting everybody straightened out. They live in a world of careless and sloppy people who never get it quite right and consequently are a threat and a burden to them. It is very frustrating to have everyone dropping the ball on you all the time. However it does allow you to be better than the rest and rescue them. You can sit back on high and complain bitterly about the incompetence of others or ride to their rescue with kind condescension.

Perfectionists are never happy with others or themselves. They are compelled to constantly monitor the details and make sure things get done just right. Unfortunately there is only one way to do anything and that is the best way, their way. They can easily degenerate into a "my way or the highway" stance with others. They often avoid arguments because they would end up arguing with everyone. In a powerful position they are overbearing and intrusive, constantly violating the boundaries of others to set them straight and make things right. They live in constant fear of making a mistake in public and will go to any length to prove themselves right if any doubts arise. Because they avoid mistakes and feel others will think less of them. They tend to be rigid in their ways and slow to innovate when confronted with problems. Their lack of flexibility may make them seem stubborn to others and is a liability in situations demanding constant reassessment and adaptation.

Having developed a foolproof mechanism for protecting themselves from criticism, they are on a relentless crusade to convince others of their perfection. Usually a deep-seated low self-esteem born of overcritical parents is the true motivation behind their behavior. Men who are perfectionistic tend to take it out angrily on others while women tend to become depressed.

Flexible

This characteristic is often the hallmark of relaxed and patient individuals. They enjoy sharing ideas and exploring points of view. They tend not to worry too much about details and innovate solutions as they go along. They tend to be trusting and forgiving. They are not afraid to make mistakes and explore new territory. They like to try out new things. They see every problem as having many possible solutions and are tolerant of the different approaches others develop toward any given situation. They are quick to forgive others and let things go. They tend to be emotionally intelligent and sympathetic to the needs of others. They tend to discuss things rather than argue because they have limited agendas and fewer expectations. They are usually very friendly and easy to get along with. They are more likely to be open to their own mistakes and to apologize for making them.

In the extreme they can be dependent and lazy. They may put things off too long and become sloppy with their work. They may lack direction and get lost in constant socializing. It may be difficult for them to make decisions and they may become involved in too many projects that never get completed. They may have a hard time saying no and defending their boundaries.

Assertive

Assertive people are usually open and direct. They respond quickly and enthusiastically to communication. They are quick to state their position and clarify the details. They are not afraid to ask questions and focus on details others may disregard. Their goal is to establish their boundaries and become clear regarding the boundaries of others. They want to know the rules of the game wherever they go so that they can employ them appropriately to avoid unnecessary confusion. They are proactive and like to anticipate and defuse problems early. They do not shy away from conflict but are not pushy or overbearing. They have a strong sense of self-efficacy and a fairly well grounded self-esteem. They are not easily manipulated and are quick to say "no" when confronted with a situation they feel uncomfortable in.

Passivity

Passive people tend to be afraid of any form of friction or conflict. They avoid confrontations at any cost. Consequently they constantly hide their feelings. They are very afraid of hurting the feelings of others and often suffer considerable guilt for every minor infraction regarding others. They are very unclear on their own boundaries because they violate them by saying "yes" so often, even though they would rather say "no". They are confused about who they are. They are usually angry because they constantly feel that others are walking all over them. They feel unseen and unheard. They occasionally have outbursts of anger when their boundaries have been violated too much and then feel deeply ashamed of their behavior. They do not like talking to people in authority or strangers. When service is poor they rarely confront the individual or complain to the management. They are unclear regarding their feelings because they hide them so often.

Impulsivity

This dimension of behavior is very grounded in a person's neurophysiology. We know from research that impulsivity is usually the result of too much slow wave activity in the frontal cortex. The term AD/HD is often applied to individuals with too much impulsivity. These are the people who color outside the lines without thinking. Impulsive people have great difficulty holding back their urges and impulses. It is hard for them to sit still and concentrate. They often loose focus unless they are engaged in a very rewarding task. Their attention wanders constantly and they tend to lose important information from discussions, lectures, and even conversations. This makes them appear careless, rude, unconcerned and irresponsible. They are often poorly organized and constantly lose things. They rush through projects and boring tasks. They frequently do not finish what they start. They may say things without weighing the consequences and act on the spur of the moment with little planning. They avoid details that are often important. They are driven by the excitement of the moment rather than careful assessment and consideration of consequences. Their emotions often overwhelm them and it may be hard for them to calm themselves after an exciting or emotion laden event. Because of these features of behavior they often break rules, norms, and mores that annoy, frustrate, and aggravate others. As a consequence they may continually receive complaints, admonishments, and even threats from others, especially individuals in positions of authority. This may lead them to see the world as an angry and dangerous place. They may become rebels without a cause. They are often anxious because they are constantly concerned with who is going to be angry or frustrated with them next. They are constantly dealing with their own guilt and the annoyance of others. They may develop a victim mentality and feel persecuted. This can lead to situations where they feel they can never win or achieve anything, at which point depression may also set in.

Regulated

These people are focused and methodical. They exhibit considerable self-discipline in most areas of their life. They take things one step at a time and pay attention to detail. They cultivate good habits in a thoughtful manner. They plan ahead and think things through carefully before acting. They take their time and make sure they get things right. They are slow to anger and calm down from an emotional upset fairly quickly. They don't procrastinate and are often well prepared in advance for unexpected events. They are well organized and rarely overwhelmed. They tend to avoid excess in most things. They make out well ordered and realistic "to do lists" and are consistent in their follow through. They appear relaxed and set aside personal time to take good care of their own needs. Because of their consistency and the success it has brought them they are confident and don't worry excessively. They have good sleep habits and carefully watch what they eat and get plenty of exercise. They like compromise and expect it. They are careful and considerate in their communications. They are patient with disruptions and upsets of others and are willing to go the extra mile to resolve problems and arrive at good solutions.

Inhibited

Inhibited people tend to be shy and anxious. They worry a lot. A great deal of their problem may be genetic and based in temperament at birth. With the right support they can overcome their inhibition, but more often than not their problem is misunderstood. They tend to fall between the cracks at home and at school. They frequently become invisible and unnoticed. They avoid drawing attention to themselves and when they get it they tend to fumble and stammer. They have a long history of unsuccessful communication and embarrassing moments that discourage them even more from asserting themselves. They are tense and uncertain. They are usually overly self-conscious and constantly worry what others might think of them. They may ruminate excessively about how they performed even in the most innocent of situations. They are never sure what to say to others. They freeze up in social situations and consequently may avoid them as much as possible. They usually feel inadequate and overwhelmed. Because of their inhibition, they often respond to people with silence and a poker face. This tends to put others off and gets misread as arrogance or "being stuck up." Others may see them as cool, aloof, and unfriendly. Inhibited people are tentative and conservative in their actions. They dislike taking risks in social situations. They may latch on to others who have more confidence and hide behind them.

Relaxed

Individuals who are relaxed tend to be open and free with their feelings and communicate them readily. They are comfortable with self-disclosures and make friends easily. They tend to trust others and value the input that others have for them. They are not afraid of conflict or friction between people and tend to recover quickly from confrontations. They plan things well but leave room for the unexpected. They are very creative when mistakes and exceptions occur and are often able to turn them into positive outcomes. They have a "live and let live" approach to life. Having fun comes easily to them and they usually have a healthy sense of humor. They are confident and tend to worry only when important issues arise that have proven to be adverse in the past. When facing adversity they are usually patient and consistent in their efforts to overcome it. When they encounter disappointment they are usually able to let go of their agenda without great anger and turmoil. They tend to be centered and accepting of their circumstance. They are quick to forgive others. They trust life in general and are able to let go and go along for the ride as much as they enjoy directing the action.

10

Take The ISI

It is part of the cure to wish to be cured.
Seneca

This is the part where you actually take action to change.

Interactive Self Inventory

Rate each statement on a scale of 1-5 according to the following key:
1= Never   2= Sometimes   3= Often   4= Usually   5= Always

1	2	3	4	5	I avoid talking to people unless I know them well.
1	2	3	4	5	I am nervous with people unless I know them well.
1	2	3	4	5	Being introduced to people makes me tense and nervous.
1	2	3	4	5	I avoid walking up and joining a group of people.
1	2	3	4	5	I feel on edge when I am with a group of people.
1	2	3	4	5	I think up excuses in order to avoid social engagements.
1	2	3	4	5	I jump at the chance to meet new people.
1	2	3	4	5	I am relaxed when I am with a group of people.
1	2	3	4	5	I don't mind talking to people at parties or social gatherings.
1	2	3	4	5	I find it easy to relax with other people.
1	2	3	4	5	I look forward to my social engagements.
1	2	3	4	5	I enjoy introducing people to each other.
1	2	3	4	5	I do my best work when I know it will be appreciated.
1	2	3	4	5	Disapproval of someone I care about is very painful for me.
1	2	3	4	5	I easily get discouraged when I don't get what I need from others.
1	2	3	4	5	I must have one person who is very special to me.

1	2	3	4	5	I need to have one person who puts me above all others.
1	2	3	4	5	I enjoy being by myself.
1	2	3	4	5	I don't depend on other people to make me feel good.
1	2	3	4	5	I like to rely on myself to get things done.
1	2	3	4	5	I don't need much from people.
1	2	3	4	5	What people think of me doesn't affect how I feel.
1	2	3	4	5	What other people say doesn't bother me.
1	2	3	4	5	Winning in competition makes me feel more powerful.
1	2	3	4	5	I become competitive even in situations that do not call for competition.
1	2	3	4	5	I tend to turn a friendly game or activity into a serious contest.
1	2	3	4	5	I feel envy when my competitors get rewarded.
1	2	3	4	5	I am very unhappy when I lose in competition.
1	2	3	4	5	I can't stand to lose an argument.
1	2	3	4	5	People who quit during competition are weak.
1	2	3	4	5	Competition inspires me to excel.
1	2	3	4	5	I get useful information from others.
1	2	3	4	5	When a person gets upset they should talk it over with others.
1	2	3	4	5	Its important to ask for help when you need it.
1	2	3	4	5	Its fairly easy to tell who you can trust and who you can't.
1	2	3	4	5	I enjoy working with others to solve a problem.
1	2	3	4	5	Two heads are better than one.
1	2	3	4	5	I hate being less than the best at things.
1	2	3	4	5	People will probably think less of me if I make a mistake.
1	2	3	4	5	People take advantage of your mistakes.
1	2	3	4	5	There is right way and a wrong way to do things.
1	2	3	4	5	I pay very close attention to details.
1	2	3	4	5	I make sure I don't make the same mistake twice.
1	2	3	4	5	Either I do something well, or I don't do it at all.
1	2	3	4	5	I make sure there are no loose ends.
1	2	3	4	5	I learn new things when I make mistakes.

1	2	3	4	5	Its good to hear a fresh point of view.
1	2	3	4	5	I make room for the unexpected in my projects.
1	2	3	4	5	I make it up as I go along.
1	2	3	4	5	It's important to be patient with the mistakes of others.
1	2	3	4	5	There are several effective ways to achieve the same goal.
1	2	3	4	5	At times a sudden change of plans is necessary.
1	2	3	4	5	I enjoy trying out new things.
1	2	3	4	5	I am open and honest about my feelings.
1	2	3	4	5	I enjoy meeting and talking with people for the first time.
1	2	3	4	5	I complain about poor service when I encounter it.
1	2	3	4	5	When I am asked to do something, I always want to know why.
1	2	3	4	5	If I don't like something I say so.
1	2	3	4	5	I don't mind confronting others when they are out of line.
1	2	3	4	5	I have a hard time saying "No."
1	2	3	4	5	I tend not to show my feelings rather than upsetting others.
1	2	3	4	5	I avoid complaining about things I don't like.
1	2	3	4	5	People should be nice to each other no matter what happens.
1	2	3	4	5	I avoid arguments of any kind.
1	2	3	4	5	I don't asked questions for fear of sounding stupid.
1	2	3	4	5	I don't plan things as carefully as I should.
1	2	3	4	5	I find it hard to sit for long periods of time.
1	2	3	4	5	I solve problems by trial and error.
1	2	3	4	5	I am restless at lectures or talks.
1	2	3	4	5	I don't always finish what I start.
1	2	3	4	5	I act on the spur of the moment.
1	2	3	4	5	I let the details take care of themselves.
1	2	3	4	5	I say things without thinking.
1	2	3	4	5	I having difficulty staying focused.
1	2	3	4	5	I like to get the job done quickly.
1	2	3	4	5	I plan things carefully.
1	2	3	4	5	When I get angry, I wait for a while before I respond.
1	2	3	4	5	I like to think carefully before I make a decision.

1	2	3	4	5	I plan each day with a written list.
1	2	3	4	5	I avoid excess in anything.
1	2	3	4	5	I set aside time for myself each day.
1	2	3	4	5	I become very upset when things don't go my way.
1	2	3	4	5	I tend to keep things well organized.
1	2	3	4	5	My worries overwhelm me.
1	2	3	4	5	I feel tense and uncertain.
1	2	3	4	5	I wish I could just be myself.
1	2	3	4	5	I am concerned about what others think.
1	2	3	4	5	I freeze up in certain situations.
1	2	3	4	5	I feel less than adequate.
1	2	3	4	5	I'm afraid of sounding stupid.
1	2	3	4	5	I'm not sure what to say.
1	2	3	4	5	I get embarrassed very easily.
1	2	3	4	5	I worry about making a good impression.
1	2	3	4	5	I can usually solve most problems if I just keep trying.
1	2	3	4	5	Things work out okay.
1	2	3	4	5	Generally I trust people to do what's right.
1	2	3	4	5	I get along well with people in general.
1	2	3	4	5	A little friction between people is normal.
1	2	3	4	5	I subscribe to the idea of live and let live.
1	2	3	4	5	I don't get hung up on things for long.
1	2	3	4	5	I like to take my time and do things right.
1	2	3	4	5	I feel tired and low energy.
1	2	3	4	5	It's very hard to get motivated.
1	2	3	4	5	Things look hopeless.
1	2	3	4	5	I feel very sad these days.
1	2	3	4	5	I feel worthless or guilty.
1	2	3	4	5	I feel like a failure
1	2	3	4	5	I seem to cry for little or no reason.
1	2	3	4	5	I am sleeping in a lot.
1	2	3	4	5	I am more irritable than usual.
1	2	3	4	5	I am eating more or less than usual.
1	2	3	4	5	I have lost interest in things that I used to enjoy.
1	2	3	4	5	I have a lot more negative thoughts about myself.

1	2	3	4	5	I have trouble going to sleep or staying asleep.
1	2	3	4	5	I feel restless and agitated.
1	2	3	4	5	I worry constantly.
1	2	3	4	5	Things feel out of control.
1	2	3	4	5	My heart feels like it is racing or pounding.
1	2	3	4	5	I feel lightheaded, faint, or dizzy.
1	2	3	4	5	My hands shake.
1	2	3	4	5	I feel short of breath or tight in the chest.
1	2	3	4	5	I sweat a lot.
1	2	3	4	5	My hands or feet feel cold.

The Automatic Self: References

Alpert, Richard: Ram Dass (1976). Be Here Now. New York. Crown Publishing Group.

Austin, James H. (1998). Zen and the Brain: Toward an Understanding of Meditation and Consciousness. Cambridge, MA: The MIT Press.

Baehr, E., Rosenfeld, J.P., & R. Baehr (1997). The Clinical Use of An Alpha Asymmetry Protocol in the Neurofeedback Treatment of Depression: Two Case Studies. Journal of Neurotherapy, 3, 12-23.

Banquet, J. P. (1973). Spectral analysis of the EEG in meditation. Electroencephalography and Clinical Neurophysiology 35: 143-151.

Beck, A. T. (1979). Cognitive Therapy and the Emotional Disorders. Cleveland, OH: Meridian.

Beck, Judith S. (1995). Cognitive Therapy: Basic and Beyond. New York: The Guilford Press.

Blum, K., Cull, J., Braverman. E., and Comings, D., (1996). Reward deficiency syndrome. American Scientist, 84, 132-145.

Bozarth, Michael A. (1991). The Mesolimbic Dopamine System as a Model Reward System. In Paul Willner & Jorgen Scheel-Kruger (Eds.), The Mesolimbic Dopamine System: From Motivation to Action. New York: John Wiley & Sons.

Cade, Maxwell and Nona Coxhead (1987). The Awakened Mind. Element Books: Dorset, England

Changeux, Jean-Pierre (1985). Neuronal Man: The Biology of Mind. Princton: Princton University Press.

Cozolino, Louis (2002). The Neuroscience of Psychotherapy: Building and Rebuilding The Human Brain. New York: Norton.

Csikszentmihalyi, Mihaly (1991). Flow: The Psychology of Optimal Performance. New York: Harper Collins.

Davidson, Richard J. (2000). Affective Style, Psychopathology, and Resilience: Brain Mechanisms and Plasticity, American Psychologist.

Davidson, R.J., Jackson,D.C., and Kalin, N.H. (2000). Emotion, Plasticity, Context, and Regulation: Perspectives From Affective Neuroscience. Psychological Bulletin, vol.126, no. 6, 890-909.

Davidson, R.J., Kabat-Zinn, J., Schumacher, J., Rosenkranz, M, Muller, D., Santorelli, S., Urbanowski, F., Harrington, A., Bonus, K., Sheridan, J. (2003). Alterations and Immune Function Produced by Mindfullness Meditation. Psychosomatic Medicine 65: 564-570.

DeLuca, J.W and Daly, R. (2003). The inner alchemy of Buddhist Tantric meditation: A qEEG case study using Low resonance, Electromagnetic Topography (LORETA). Subtle Energies & Energy Medicine, 13 (2), pp 1-54.

Demasio, Antonio (1994). Descates' Error: Emotion, Reason, and the Human Brain. New York: Avon Books.

Demasio, Antonio (1999). The Feeling of What Happens: Body and Emotion in the Making of Consciousness. New York: Harcourt Brace & Company.

Dinkmeyer, D. & McKay, G. (1976). Systematic Training for Effective Parenting. Circle Pines, MN: American Guidance Service.

Drevets, W.C., Price, J.L.. Simpson, J.R.Jr, et al: Subgenual prefrontal cortex abnormalities in mood disorders. Nature 386:824-827, 1997.

Dunn, Bruce R., Hartigan, Judith A., & William L. Mikulas (1999). Concentration and Mindfulness Meditations: Unique Forms of Consciousness? Applied Psychophysiology and Biofeedback, Vol. 24, No. 3, 147-165.

Duffy,F., Hughes, J.R.,Miranda, F.,Bernaqd, P., and Cook, P. (1994). Status of quantitative EEG (QEEG) in clinical practice. Clinical EEG,25(4), VI-XXII.

Egner T, Gruzelier, J.H. (2003). Ecological validity of neurofeedback: modulation of slow wave EEG enhances musical performance. Neuroreport. 14: 1221-1224.

Engler, Jack, Brown, Daniel P., Wilber, Ken (1986). Transformations of Consciousness: Conventional and Contemplative Perspectives on Development. Shambala:Boston.

Erikson, Erik H. (1968). Childhood and Society. New York: Norton.

Femi, L. (1998). Open Focus Attention. A Futurehealth Workshop. Palm Springs, CA.

Fehmi, L.G. (1978). EEG biofeedback, multichannel synchrony training, and attention. Chapter in A.A. Sugarman & R.E. Tarter (Eds.), Expanding Dimesnions of Consciusness. New York: Springer.

Fehmi, L.G. & Selzer, F.A. (1980). Biofeedback and attention training. Chapter in S. Boorstein (Ed.), Transpersonal Psychotherapy. Palo Alto: Science and Behavior Books.

Goleman, Daniel (1995). Emotional Intelligence. New York: Bantam Books.

Goleman, Daniel. (1988). The Meditative Mind. New York: G. P. Putnam's Sons.

Goleman, Daniel (2003). Destructive Emotions: How Can We Overcome Them? New York: Bantam Books.

Goswami, Amit (2000). The Visionary Window. Wheaton, Ill.: The Theosophical Publishing House.

Green, E. E. & Green, A. M. (1977). Beyond Biofeedback. San Francisco: Delacarte.

Jaynes, Julian (1976). The Origin of Consciousness in the Breakdown of the Bicameral Mind. Boston: Houghton Mifflin.

Hawking, S. (1988). A Brief History of Time: From the Big Bang to Black Holes. New York: Bantam.

Hughes, John R., and John, E. Roy (1999). Conventional and Quantitative Electroencephalography in Psychiatry. Journal of Neuropsychiatry and Clinical Neuroscience, 11:2, Spring.

Hughes, Stuart W. and Crunelli, Vincenzo (2005). Thalamic mechanisms of EEG alpha rhythms and their pathological implications. The Neuroscientist, vol. 11, number 4.

Isherwood, Christopher & Swami Prabhavananda (1969). How To Know God: The Yoga Aphorisms Of Patanjali. New York: Mentor.

Kamalashila (1995). Meditation: The Buddhist Way of Tranquility and Insight. Surrey, England: Biddles Ltd.

Kamiya, J. (1969). Operant control of the EEG alpha rhythms and some of its reported effect on consciousness. In: Tart, C.T. (Ed) Altered States of Consciousness, p573, New York: Wiley & Sons.

Kaplan, G. B., and Hammer, R.P. (2002). Brain Circuitry and Signaling in Psychiatry: Basic Science and Implications. American Psychiatric Publishing: Washington, DC.

Kasamatsu, A., and Hirai, T. An EEG Study on the Zen Meditation." In C. Tart (Ed.), Altered States of Consiousness. New York: Wiley, 1969.

Koch, Christof (2004). The Quest for Consciousness: A Neurobiological Approach. Roberts and Company: Englewood, Colorado.

Kruger (Eds.), The Mesolimbic Dopamine System: From Motivation to Action. New York: John Wiley & Sons.

Kuhn, Thomas S. (1970). The Structure of Scientific Revolutions. Chicago: The University of Chicago Press.

Le Doux, Joseph. (1996). The Emotional Brain: The Mysterious Underpinnings of Emotional Life. New York: Simon & Schuster.

Le Doux, Joseph (2002). The Synaptic Self: How Our Brains Became Who We Are. Viking: New York.

McEwen, Bruce (1987). Influence of Hormones and Neuroactive Substances on Immune Function. In Carl W. Cotman (Ed.), The Neuro-Immune-Endocrine Connection. New York: Raven Press.

McCormick, David A. (1999). Are thalamocortical rhythms the rosetta stone of a subset of neurological disorders? Nature Medicine, vol. 5, number 12.

Maas, James B. (1999). Power Sleep. New York: Collins.

Mindcycles, 2004. This research with brainmaps is published on the website www.mindcycles.com.

Niedermeyer, Ernst, Lopes Da Silva, Fernando (1999). Electroencephalography: Basic Principles, Clinical Applications, and Related Fields. Fourth Edition. Lippincott Williams & Eilkins: New York

Newberg, Andrew (2001). Why God Won't Go Away. New York: Ballantine Books.

Nunez, Paul L. (Ed.) . (1995). Neocortical Dynamics and Human EEG Rhythms. New York: Oxford University Press.

Ornstein, Robert E & Claudio Naranjo (1974). On the Psychology of Meditation. New York: Viking Press.

Pribram, Karl H. (1977). Languages of the Brain: Experimental Paradoxes and Principles In Neuropsychology. Monterey, CA: Cole Publishing.

John, E. R., Prichep, L. S., Fridman, J., & Easton, P. (1988). Neurometrics: Computer Assisted Differential Diagnosis of Brain Dysfunctions. Science 293, 162-169.

Peniston, B. G., & Kulkosky, P.J. (1989) Alpha-theta Brainwave Training and Beta-Endorphin Levels in Alcoholics. Alcohol. Clin. Exp. Res. 13, 271-279.

Posner, M.I. & Raichle, M.E. (1997). Images of Mind. New York: Scientific American.

Quine, W.V.O. (1953). From a Logical Point of View. New York. Harper & Row.

Robbins, Jim (2000). A Symphony In The Brain. New York. Grove Hills.

Ruden, Ronald A. (1997). The Craving Brain: The Biobalance Approach to Controlling Addictions. New York: HarperCollins.

Rudin, Ron (1999). Personal conversations with Ron while writing Mindfitness with Adam Crane.

Satinover, Jeffrey (2001). The Quantum Brain. New York:John Wiley& Sons.

Szazs, Thomas S. (1984). The Myth of Mental Illness. New York: Perennial Library Harper.

Schwartz, Jeffrey M. and Sharon Begley (2002). The Mind and The Brain. Harper Collins: New York.

Schore, A. N. (1994). Affect Regulation and the Origin of the Self: The Neurobiology of Emotional Development. Hillsdale, NJ: Lawrenece Erlbaum Associates.

Soutar, R & Crane, A. (2000). Mindfitness Training: Neurofeedback and the Process. i Universe: New York..

Sterman, M.B., & Friar L. (1972). Suppression of seizures in epileptic following sensorimotor EEG biofeedback training. EEG Clinical Neurophysiology. 33, 89-95.

Sterman, M. B. (1986). Epilepsy and its treatment with EEG feedback therapy. Annals of Behavioral Medicine, 8, 21-25.

Sterman, M.B. & C.A. Mann (1995). Concepts and Applications of EEG Analysis in Aviation Performance Evaluation. Biological Psychology, 40, 115-130.

Talbot, Michael (1991). The Holographic Universe. Harper Collins: New York.

Thatcher, R. W. (1998). Normtive EEG databases and EEG biofeedback. Journal of Neurotherapy.2(4),8-39.

Wallace, R.K. (1970). Physiological effects of Transcendental Meditation. Science, 167, pp1751-1754.

Westcott, M. (1973). Hemispheric symmetry of the EEG during the Transcendental Meditation technique. Department of Psychology, University of Durham, Durham, England, 1973.

Watts, A. (1961). Psychotherapy East and West. New York: Pantheon.

Wise, Anna (1997). The High Performance Mind. New York: G.P. Putnam's Sons.

Wolf, Fred Alan. (1994). The Dreaming Universe. New York: Simon & Schuster.

Wolinsky, Stephen (1991). Trances People Live. Conneticut: The Bramble Company

978-0-595-38695-6
0-595-38695-4

Printed in the United States
57785LVS00005B/403-414